# Orbital Mechanics: The Essentials

David A. Cicci

ISBN-13: 979-8-9866183-2-6 (Paperback)

Cover design by Darby A. Cicci

Cover image: Webb's First Deep Field shows galaxy cluster SMAC 0723, provided courtesy of: NASA, ESA, CSA, STScI.



DS Publishing
Auburn, AL

Printed in the United States of America.

# PREFACE

During my 30-plus years of teaching orbital mechanics in the Department of Aerospace Engineering at Auburn University, I carefully developed the material I taught. I intended to cover the introductory topics that were essential for aerospace engineering students. The material I developed during my teaching career is now offered in *Orbital Mechanics: The Essentials.* I believe it's presented in a simplified yet thorough manner, which facilitates understanding for students as well as working professionals. The topics are covered on a comprehensive level that also provides a solid background for students in other fields of engineering and science who might be interested in learning orbital mechanics. My description of the topics assumes readers have an introductory knowledge of differential and integral calculus and differential equations. However, individuals without such a background can still benefit from this textbook.

*Orbital Mechanics: The Essentials* comprising an in-depth discussion to the two-body problem and an introduction to satellite perturbations. It includes hundreds of problems designed to improve student understanding. Special effort was taken to select problems that demonstrate applications of the concepts covered. The problems were taken from homework and examinations that thousands of Auburn aerospace engineering students worked during their time in my classes.

As an engineering professor, my primary goal was to simplify difficult technical concepts so all students could understand them. For that reason, Orbital *Mechanics: The Essentials* is organized in a module-based learning format, where each module only covers a few important concepts without a multitude of derivations and mathematical proofs. Each module also includes several example problems with solutions and a larger selection of problems with answers, which are appropriate for homework assignments or in-class exercises.

The material in this textbook is presented in more of an outline/powerpoint type of display rather than in wordy and overly detailed descriptions many books include. I've found much of the information in standard textbooks is too complex and overwhelming to undergraduate students, which discourages rather than motivates them. It's my belief that if students learn the fundamentals there are many reference books available from which they can expand their knowledge base about any specific topic. Orbital mechanics is not easy, but it's been my goal to make it enjoyably simple once the basic laws are understood. To do so, I've attempted to present the difficult concepts as clearly as possible to facilitate that understanding.

I've also always been concerned about the high cost of textbooks. I don't believe students should be required to spend hundreds of dollars for textbooks for any one course, especially when they might be enrolled in five or six courses in a semester. As a result, I've chosen to create a textbook using simple drawings and figures rather than flowery, multi-colored artwork to help keep the cost of this textbook affordable to students.

I hope all who use this textbook find it understandable and useful. I guarantee that those working through it will learn a tremendous amount about orbital mechanics.

David A. Cicci, Ph.D.
Professor Emeritus
Department of Aerospace Engineering
Auburn University
Auburn, AL
ciccida@auburn.edu

# TABLE OF CONTENTS

# ORBITAL MECHANICS: THE ESSENTIALS

(This page was intentionally left blank.)

# Orbital Mechanics: The Essentials

David A. Cicci

(This page was intentionally left blank.)

# Module 1: Coordinate System Description

There are many coordinate systems used in the field of orbital mechanics. Five of the most common coordinate systems and those that will be described in this module are

- Heliocentric-ecliptic, or the solar-ecliptic (SE), coordinate system.
- Geocentric-equatorial, or Earth-centered inertial (ECI) coordinate system.
- Equatorial, right ascension-declination, or celestial, coordinate system.
- Earth-centered, Earth-fixed (ECEF), or geographical, coordinate system.
- Azimuth-elevation, or topocentric-horizon (SEZ), coordinate system.
- Perifocal coordinate system.

Coordinate systems are classified as being either 'inertial' or 'relative'. <u>Inertial coordinate systems</u> are ones which are either fixed or move with a constant speed in a straight line, i.e., they are not accelerating.

<u>Relative coordinate systems</u> may have either a rectilinear acceleration, i.e., changing velocity, or centripetal acceleration, i.e., motion on a curved path with constant, or changing velocity.

While no coordinate system is truly inertial, the SE and ECI systems can be considered to be inertial because their accelerations are very small, while the others listed above are generally considered to be relative.

Many other coordinate systems are used in orbital mechanics, which, along with the transformations between them, are described in Ref. [1].

All physical constants and astronomical data required for this text are provided in Appendix A.

<u>Heliocentric-Ecliptic, or Solar-Ecliptic (SE), Coordinate System</u>

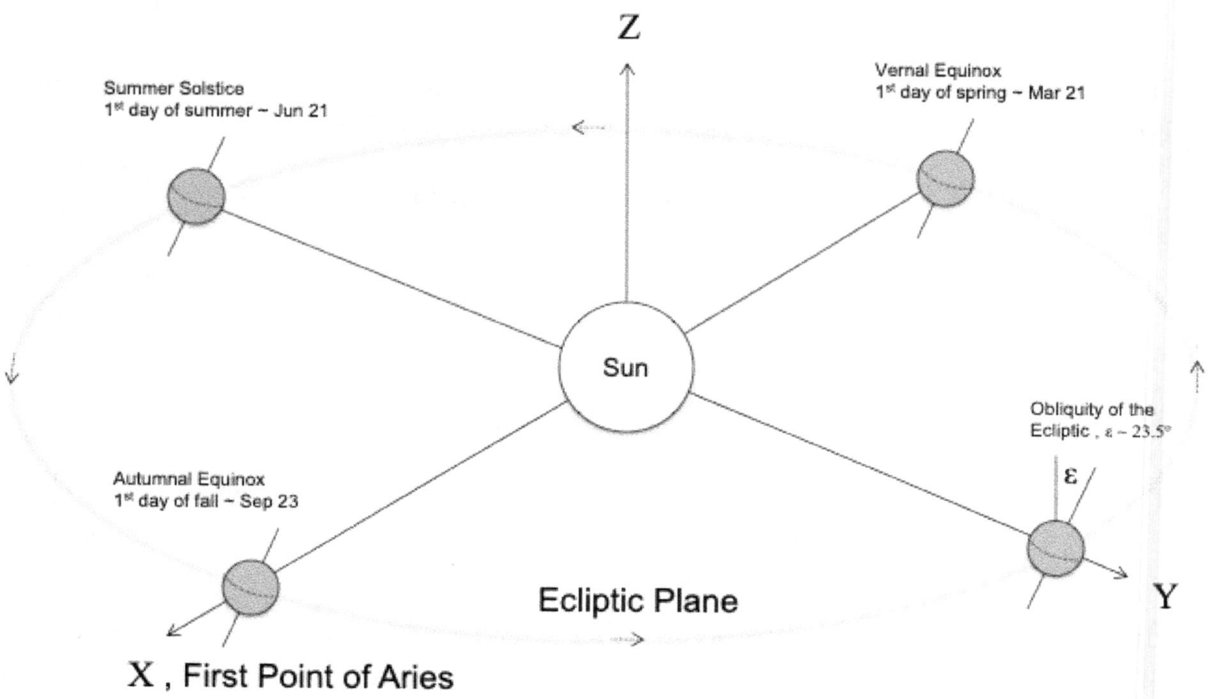


- Origin is located at the center of the Sun.

- X-axis points in the direction of the vernal equinox, i.e., First Point of Aries (which now points to a location in the constellation Pisces).

- Z-axis coincides with the Sun's North Pole.

- XY plane coincides with the plane of Earth's orbit about the Sun, i.e., the ecliptic plane.

- Earth's mean distance from the Sun, i.e., semi-major axis of its orbit, $a = 1.496 \times 10^6$ km, and the eccentricity, i.e., shape, of Earth's orbit, $e = 0.0167$.

- The point of Earth's closest approach to the Sun, or perihelion, is $r_p = 1.471 \times 10^6$ km.

- Earth's farthest point from the Sun, or aphelion, is $r_a = 1.521 \times 10^6$ km.

- Earth's orbital period, $T = 365.256$ days.

- Obliquity of the ecliptic, i.e., the inclination of the Earth's equatorial plane to the ecliptic plane, $\varepsilon = 23.442405°$.

Geocentric-Equatorial, or Earth-Centered Inertial (ECI), Coordinate System

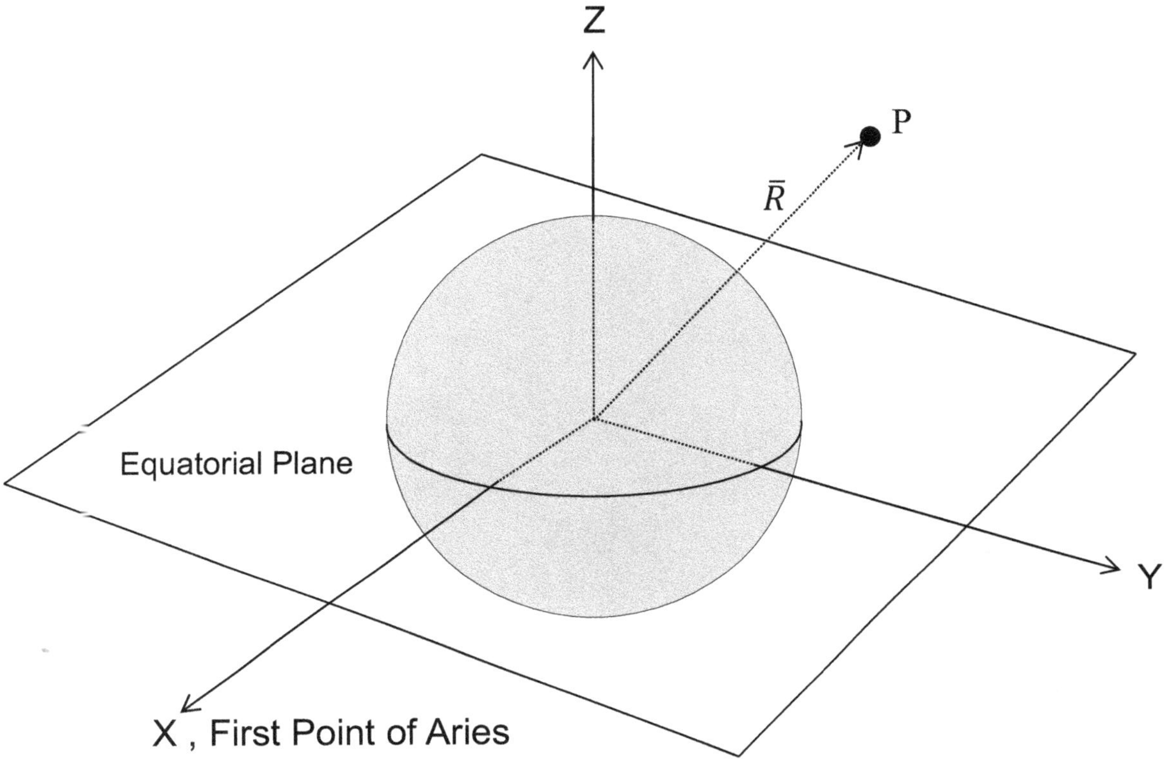


- Origin is located at the center of Earth.
- X-axis points in the direction of the vernal equinox.
- Z-axis points along the mean rotation axis of Earth.
- XY plane coincides with Earth's equatorial plane.
- $\bar{I}$, $\bar{J}$, and $\bar{K}$ are unit vectors in the X, Y, and Z directions, respectively.
- J2000 is the most common ECI coordinate system currently being used, with an epoch of noon on 1 Jan 2000.
- Earth's mean equatorial radius, $R_E = 6{,}378.1$ km.
- Point P is located by position vector $\bar{R} = X\bar{I} + Y\bar{J} + Z\bar{K}$, where $R = |\bar{R}| =$ range.
- All celestial bodies are considered to be perfectly spherical for calculation purposes in this text.

Equatorial, Right Ascension-Declination, or Celestial, Coordinate System

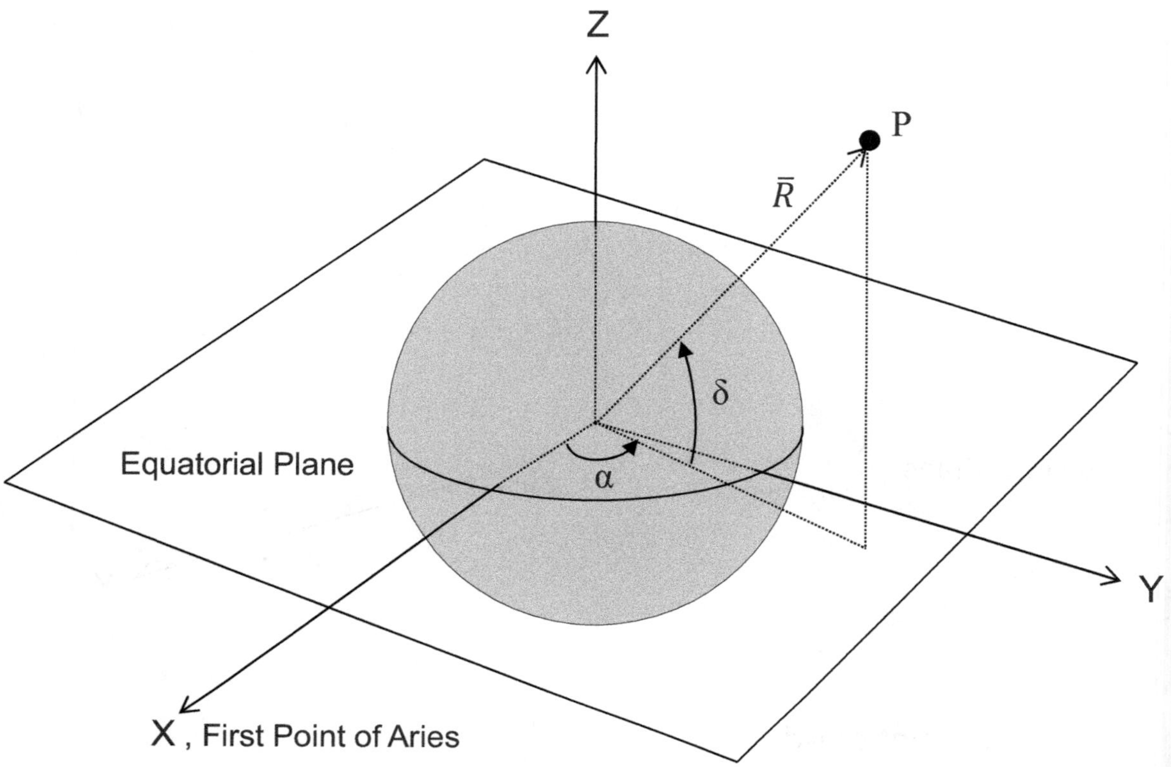


- Origin located at the center of Earth.

- X-axis points in the direction of the vernal equinox.

- Z-axis points along the mean rotation axis of Earth.

- XY plane coincides with Earth's equatorial plane.

- Point P is located by $\bar{R} = R\cos\delta\cos\alpha\,\bar{I} + R\cos\delta\sin\alpha\,\bar{J} + R\sin\delta\,\bar{K}$.

  $R = |\bar{R}|$ = range
  $\alpha$ = right ascension, measured CCW about the Z-axis from the X-axis to the projection of $\bar{R}$ in the XY plane, $0^o \le \alpha \le 360^o$
  $\delta$ = declination, measured up or down from the XY plane to $\bar{R}$, $-90^o \le \delta \le 90^o$

Earth Centered, Earth-Fixed (ECEF), or Geographical, Coordinate System

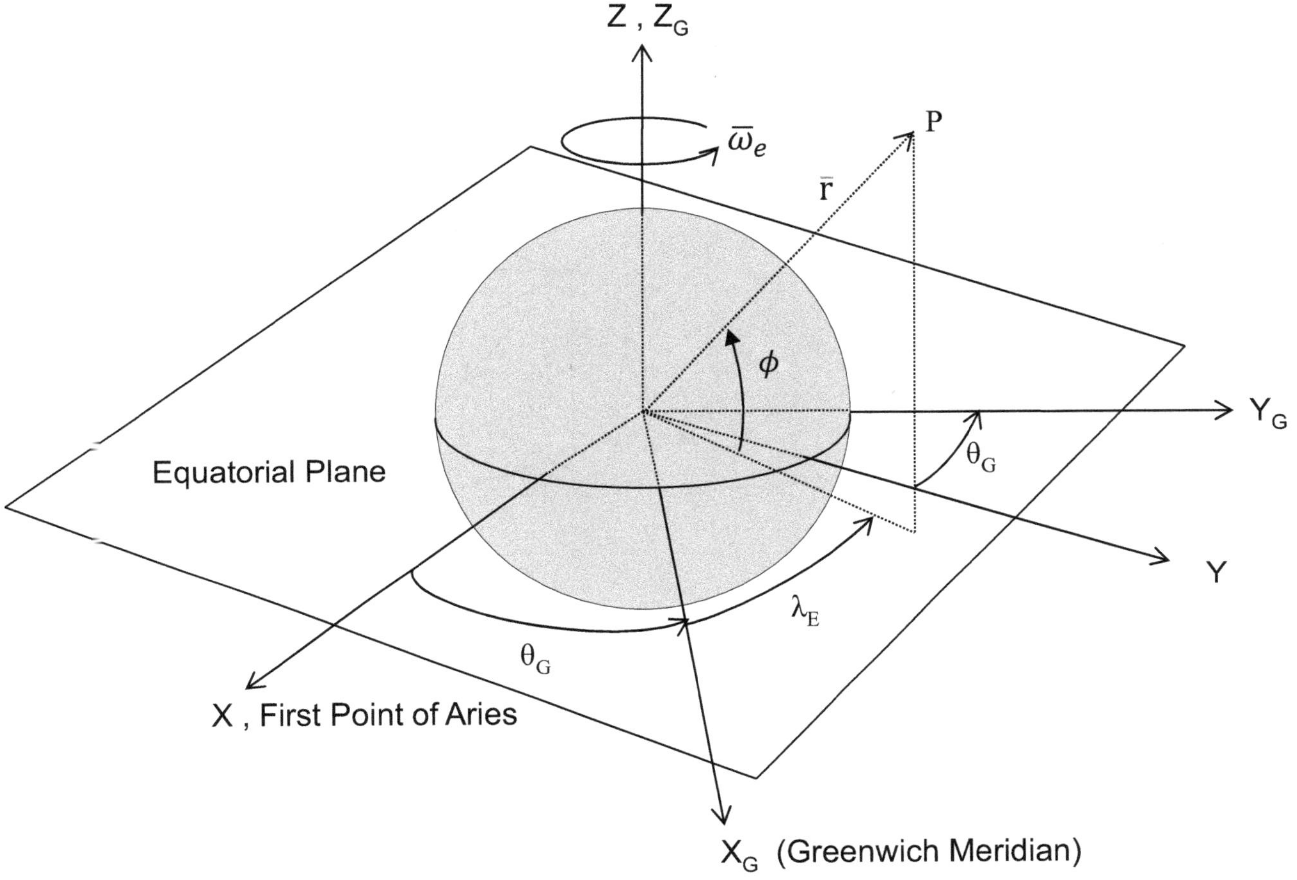


- Origin located at the center of Earth.

- $X_G$-axis passes through the Greenwich Meridian (0° longitude), having unit vector $\overline{I}_G$, and the unit vector along the $Y_G$-axis is $\overline{J}_G$.

- $Z_G$-axis points through the North Pole (coincides with ECI Z-axis), having unit vector $\overline{K}_G$.

- $X_G Y_G$ plane coincides with Earth's equatorial plane, and the $X_G$- and $Y_G$-axes rotate with the same angular velocity as Earth.

- Point P is located by $\overline{r} = r\cos\phi\cos\lambda_E\,\overline{\imath}_G + r\cos\phi\sin\lambda_E\,\overline{\jmath}_G + r\sin\phi\,\overline{k}_G$.

  $r = |\overline{r}| =$ range
  $\lambda_E =$ East Longitude, measured CCW about the $Z_G$-axis from the Greenwich Meridian, $0^o \le \lambda_E \le 360^o$ and $\lambda_W =$ West Longitude $= 360^o - \lambda_E$
  $\phi =$ latitude, measured up or down from the XY plane to $\overline{r}$, $-90^o \le \phi \le 90^o$

Azimuth-Elevation, or Topocentric-Horizon (SEZ), Coordinate System

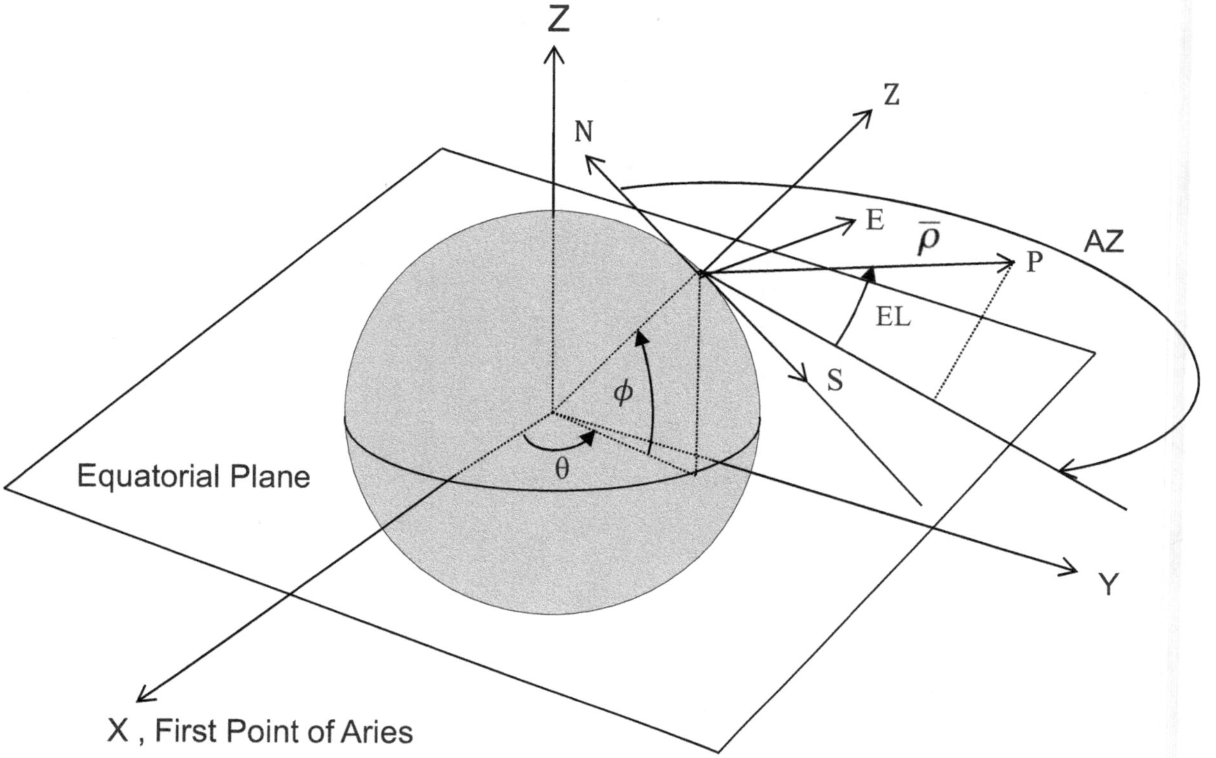


- Origin located at a body-fixed site on Earth's surface, e.g., tracking station.
- S-axis points due south, E-axis points due east, and Z-axis points radially outward along the local vertical (geographical latitude).
- SE plane is tangent to Earth at the origin.
- $\bar{s}$, $\bar{e}$, and $\bar{z}$ are unit vectors in the S, E, and Z directions, respectively.
- Point P is located by $\bar{\rho} = -\rho \cos EL \cos AZ\, \bar{s} + \rho \cos EL \sin AZ\, \bar{e} + \rho \sin EL\, \bar{z}$.

  $\rho = |\bar{\rho}| =$ range
  AZ = azimuth, measured CW about Z-axis from north to the projection of $\bar{\rho}$ in SE plane, $0^o \le AZ \le 360^o$
  EL = elevation, measured up or down from the SE plane to $\bar{\rho}$, $0^o \le EL \le 90^o$

## Perifocal Coordinate System

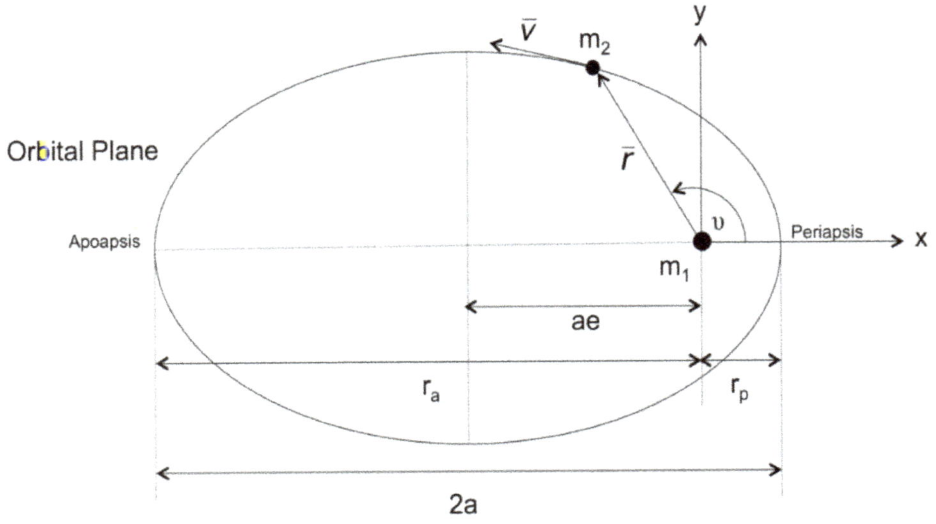


- Origin located at the center of the primary mass, $m_1$.
- x-axis points from $m_1$ through periapsis/pericenter, the point closest to $m_1$, where $r_p = a(1 - e)$
- xy plane coincides with the orbital plane of $m_2$ about $m_1$.
- z-axis is normal to the xy plane.
- $\bar{\imath}$, $\bar{\jmath}$, and $\bar{k}$ are unit vectors in the x, y, and z directions, respectively.
- Apoapsis/apocenter is the point farthest from $m_1$, where $r_a = a(1 + e)$
- a is the semi-major axis, or size, of the orbit.
- e is the eccentricity, or shape, of the orbit.
- ν is the true anomaly, the angle measured from periapsis/pericenter to $\bar{r}$ in the direction of travel of $m_2$.
- $m_2$ is located in the orbital plane by $\bar{r}$ as $\bar{r} = r\cos\nu\,\bar{\imath} + r\sin\nu\,\bar{\jmath}$.

Example 1.1

An Earth satellite is located at an altitude of 921 km, with a right ascension of 308° and a declination of 82°. Calculate the position vector of the satellite in the geocentric-equatorial (ECI) coordinate system.

Solution:

$$R = R_E + \text{alt} = 6{,}378.0 + 921 = 7{,}299.0 \text{ km}$$

$$X = R \cos\delta \cos\alpha = 7{,}299.0 \cos 82^o \cos 308^o = 625.4 \text{ km}$$

$$Y = R \cos \sin\alpha = 7{,}299.0 \cos 82^o \sin 308^o = -800.5 \text{ km}$$

$$Z = R \sin\alpha = 7{,}299.0 \sin 308^o = 7{,}288.0 \text{ km}$$

$$\bar{R} = X\,\bar{I} + Y\,\bar{J} + Z\,\bar{K}$$

$$\Rightarrow \bar{R} = 625.4\,\bar{I} - 800.5\,\bar{J} + 7{,}288.0\,\bar{K} \text{ km}$$

Example 1.2

Determine the position vector in the geographical coordinate system of a tracking station located at $\phi = -71^o, \lambda_W = 92.0^o$

Solution:

$$X = R_E \cos\phi \cos\lambda_E = 6{,}378.0 \cos(-71^o) \cos(360^o - 92^o) = -72.5 \text{ km}$$

$$Y = R_E \cos\phi \sin\lambda_E = 6{,}378.0 \cos(-71^o) \sin(360^o - 92^o) = -2{,}075.2 \text{ km}$$

$$Z = R_E \sin\phi = 6{,}378.0 \sin(-71^o) = -6{,}030.5 \text{ km}$$

$$\bar{r} = X\,\bar{\imath}_G + Y\,\bar{\jmath}_G + Z\,\bar{k}_G$$

$$\Rightarrow \bar{r} = -72.5\,\bar{\imath}_G - 2{,}075.2\,\bar{\jmath}_G - 6{,}030.5\,\bar{k}_G \text{ km}$$

Example 1.3

A ground-based tracking station obtains the tracking data below for the position of an Earth satellite. Compute the position vector from the tracking station to the satellite in the azimuth-elevation coordinate system, where $\rho = 802$ km, $AZ_1 = 92.3°$, $EL_1 = 47.9°$.

Solution:

$$S = -\rho \cos EL \cos AZ = -802 \cos 47.9° \cos 92.3° = 21.58 \text{ km}$$

$$E = \rho \cos EL \sin AZ = 802 \cos 47.9° \sin 92.3° = 537.25 \text{ km}$$

$$Z = \rho \sin EL = 802 \sin 47.9° = 595.07 \text{ km}$$

$$\bar{\rho} = S\,\bar{s} + E\,\bar{e} + Z\,\bar{z}$$

$$\Rightarrow \bar{\rho} = 21.58\,\bar{s} + 537.25\,\bar{e} + 595.07\,\bar{z} \text{ km}$$

Problems

1.1 Given that Earth's orbit about the Sun has an eccentricity of $e = 0.0167$, calculate Earth's perihelion and aphelion distances.
(Ans. $r_p = 1.471 \times 10^6$ km; $r_a = 1.521 \times 10^6$ km)

1.2 If the moon's mean distance from Earth is 384,400 km and its eccentricity is 0.0549, determine the moon's apogee and perigee in kilometers and miles.
(Ans. $r_a = 405{,}504$ km $= 251{,}968$ mi; $r_p = 363{,}296$ km $= 225{,}742$ mi)

1.3 Compute the position vector of an Earth satellite in the geocentric-equatorial (ECI) coordinate system for the tracking data provided below.
(a) altitude = 314 km, $\alpha = 97.0°$, $\delta = 19.2°$
(b) altitude = 3,745 km, $\alpha = 332.0°$, $\delta = -59.6°$
(Ans. (a) $\bar{r} = -770.20\,\bar{I} + 6{,}272.79\,\bar{J} + 2{,}200.82\,\bar{K}$ km;
(b) $\bar{r} = 4.523.03\,\bar{I} - 2{,}404.938.21\,\bar{J} - 8{,}731.34\,\bar{K}$ km)

1.4 Find the altitude, right ascension, and declination of an Earth satellite for each of the geocentric-equatorial (ECI) position vectors provided.
(a) $\bar{r} = 9{,}200\,\bar{I} - 2{,}600\,\bar{J} + 8{,}400\,\bar{K}$ km
(b) $\bar{r} = -5{,}500\,\bar{I} + 7{,}300\,\bar{J} - 3{,}400\,\bar{K}$ km
(Ans. (a) (alt) $= 6{,}348.21$ km, $\delta = 41.30°$, $\alpha = 344.22°$;
(b) (alt) $= 3{,}373.79$ km, $\delta = -20.41°$, $\alpha = 127.81°$)

1.5 For the tracking stations located at the positions given below, calculate their position vectors in the geographical coordinate system.

(a) $\phi = 73.4^{o}, \lambda_W = 209.3^{o}$

(b) $\phi = -39.1^{o}, \lambda_E = 79.6^{o}$

(Ans. (a) $\bar{r} = -1{,}589.05\,\bar{i}_G + 891.73\,\bar{j}_G + 6{,}112.31\,\bar{k}_G$ km;
(b) $\bar{r} = 893.52\,\bar{i}_G + 4{,}868.41\,\bar{j}_G - 4{,}022.54\,\bar{k}_G$ km)

1.6 Determine the latitude and east longitude of each tracking station below, whose position vectors are given in the geographical coordinate system given below. tracking station below.

(a) $\bar{r} = 709.85\,\bar{i}_G - 3{,}165.41\,\bar{j}_G + 5{,}491.53\,\bar{k}_G$ km

(b) $\bar{r} = -1{,}590.11\,\bar{i}_G + 4{,}886.98\,\bar{j}_G - 3{,}777.52\,\bar{k}_G$ km.

(Ans. (a) $\phi = 59.43^{o}, \lambda_E = 282.64^{o}$; (b) $\phi = -36.32^{o}, \lambda_E = 108.02^{o}$)

1.7 Tracking data for the positions of two Earth satellites are provided below. Calculate the position vectors from the tracking station to the satellites in the azimuth-elevation coordinate system for each set of data.

(a) $\rho_1 = 1{,}375$ km, $AZ_1 = 102.4^{o}$, $EL_1 = 42.1^{o}$

(b) $\rho_2 = 2{,}711$ km, $AZ_2 = 273.1^{o}$, $EL_2 = 53.4^{o}$

(Ans. (a) $\bar{\rho}_1 = 219.08\,\bar{s} + 996.42\,\bar{e} + 921.84\,\bar{z}$ km;
(b) $\bar{\rho}_2 = -87.41\,\bar{s} - 1{,}614.00\,\bar{e} + 2{,}176.44\,\bar{z}$ km)

1.8 The azimuth-elevation coordinates for two Earth satellites are provided by a ground-based tracking station and are given below. Find the range, azimuth, and elevation of each satellite.

(a) $\bar{\rho}_1 = 3{,}991.80\,\bar{s} - 1{,}003.38\,\bar{e} + 2{,}475.11\,\bar{z}$ km

(b) $\bar{\rho}_2 = -4{,}919.35\,\bar{s} + 967.88\,\bar{e} + 1{,}984.29\,\bar{z}$ km

(Ans. (a) $\rho_1 = 4{,}802.86$ km, $AZ_1 = 194.11^{o}$, $EL_1 = 31.02^{o}$;
(b) $\rho_1 = 5{,}392.05$ km, $AZ_2 = 11.13^{o}$, $EL_2 = 21.59^{o}$)

# Module 2: Time and Timekeeping

There are many different measures of time and methods of timekeeping used in orbital mechanics. The most common and widely used forms discussed in this module are

- Mean Solar and Sidereal Days.
- Universal Time.
- Julian Date.
- Sidereal Time.

Many other methods of time determination and timekeeping are used in orbital mechanics and are presented in Ref. [1].

## Mean Solar and Sidereal Days

A mean solar day and a sidereal day are defined as follows

- Mean solar day = the time between the Sun being directly overhead one day until it's directly overhead the next day = 24 hr (= 86,400 sec). During this time period, Earth rotates on its axis through an angle of $360.986^{\circ}$.
- Sidereal day = the time it takes for one complete rotation of Earth on its axis = 23hr 56 min 4.09 sec (86,164.09 sec). During one sidereal day, Earth rotates on its axis through an angle of $360^{\circ}$.

The reference lines on the following figure show the amount Earth rotates during a mean solar day, while the additional amount of Earth rotation during a sidereal day is shown to be equal to the angle $\psi$.

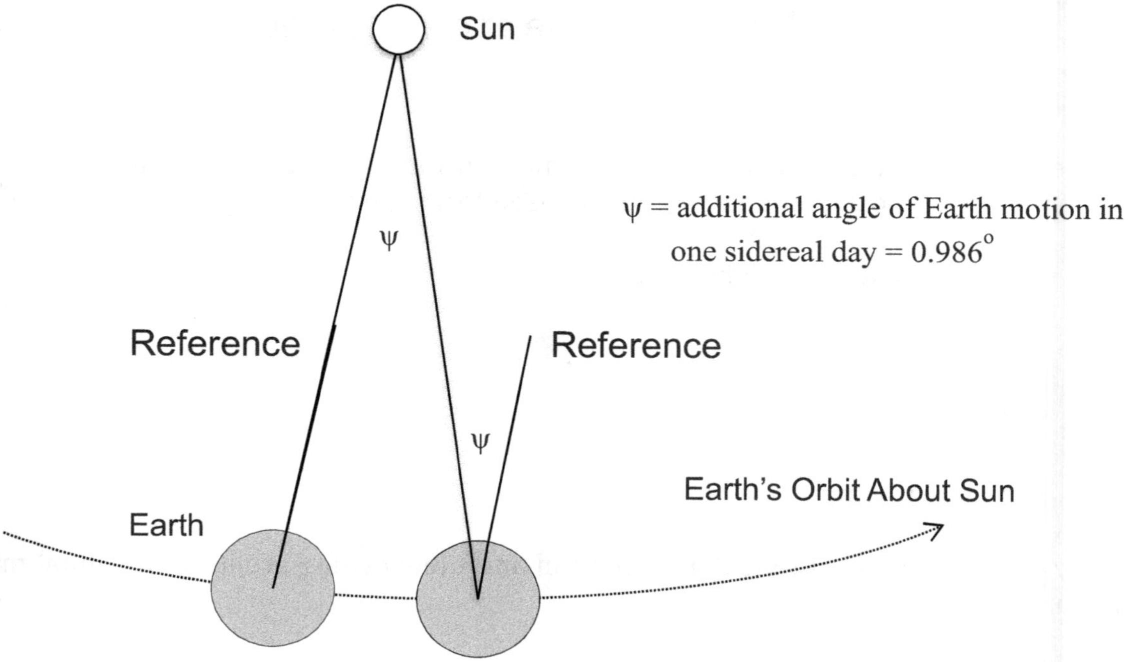


Universal Time

Universal Time (UT) is the mean solar time at the Greenwich Meridian (0° longitude).

UT is also known as Greenwich Mean Time (GMT), Coordinated Universal Time (UTC), which includes the irregular time variations due to Earth's rotation, and ZULU time, which is commonly used by the military.

Local time is UT $\pm$ 1 hour/time zone, east or west of the Greenwich Meridian.

Other commonly used time constants are

one calendar year = 365 mean solar days

one tropical year = 365.2421988 mean solar days

A tropical year is defined as the time between two successive passes of the Sun through the vernal equinox direction.

Calendar years have 365 days but have 366 days every fourth year, i.e., leap year, except in the century years whose hundreds number is not divisible by 4, e.g., the year 2000 is a leap year but the year 1900 is not a leap year.

## Julian Date

Julian Date (JD) = an arbitrary benchmark that is a continuous count of days elapsed since the chosen epoch of noon on January 1, 4713 B.C. Each Julian day is measured in whole days from noon one day to noon the next day.

The Julian Date, computed at UT, is calculated using the following formulas

$$J_0 = 367y - \mathrm{INT}\left\{\frac{7\left[y + \mathrm{INT}\left(\frac{M+9}{12}\right)\right]}{4}\right\} + \mathrm{INT}\left(\frac{275M}{9}\right) + d + 1{,}721{,}013.5 \text{ days}$$

$$JD = J_0 + \frac{h}{24} + \frac{\left(m + \frac{s}{60}\right)}{1{,}440} \text{ days}$$

where

y = year
M = month
d = day
h = hour
m = minute
s = second
INT = integer function, i.e., number truncated at the decimal point

Since the JD is measured from noon-to-noon, a JD ending in '.0' refers to noon, and a JD ending in '.5' refers to the start of the day at midnight.

Modified Julian Date (MJD) = Julian Date based on the reference epoch of midnight between the 16th and 17th of November 1858.

The MJD is also calculated at UT using the formula

$$MJD = JD - 2{,}400{,}000.5 \text{ days}$$

Example 2.1

In Greenwich, England, determine the number of seconds between the start of the day on October 19, 1921, and noon on April 12, 1980.

Solution:

$$y_1 = 1921 \;\; , \;\; M_1 = 10 \;\; , \;\; d_1 = 19 \;\; \text{at } 0^h \text{ UT}$$

$$y_2 = 1980 \;\; , \;\; M_2 = 4 \;\; , \;\; d_2 = 12 \;\; \text{at } 12^h \text{ UT}$$

$$JD_1 = 367(1921) - INT\left\{\frac{7\left[1921 + INT\left(\frac{10+9}{12}\right)\right]}{4}\right\} + INT\left[\frac{275(10)}{9}\right] + 19 + 1{,}721{,}013.5$$
$$+ 0.0 + 0.0$$

$$JD_1 = 705{,}007 - 3{,}363 + 305 + 19 + 1{,}721013.5 = 2{,}422{,}981.5 \text{ days}$$

$$JD_2 = 367(1980) - INT\left\{\frac{7\left[1980 + INT\left(\frac{4+9}{12}\right)\right]}{4}\right\} + INT\left[\frac{275(4)}{9}\right] + 12 + 1{,}721013.5$$
$$+ \frac{12}{24} + 0.0$$

$$JD_2 = 726{,}660 - 3{,}466 + 122 + 12 + 1{,}721013.5 + 0.5 = 2{,}244{,}342.0 \text{ days}$$

$$\Delta t = (JD_2 - JD_1)(86{,}400) = (2{,}244{,}342.0 - 2{,}422981.5)(86{,}400)$$

$$\Longrightarrow \;\; \Delta t = 1{,}845{,}547{,}200 \text{ sec}$$

Sidereal Time

Sidereal time is a timekeeping system used to locate celestial objects. It's based on Earth's rate of rotation relative to the fixed stars and is measured in angles, since angles can be divided in hours, minutes, and seconds. The conversions between time and angular distance in longitude are

$$24 \text{ hours} = 360^\circ \;\; \text{(midnight corresponds to } 0^\circ\text{)}$$

$$1 \text{ hour} = 15^\circ \;\; \text{(degrees longitude)}$$

$$1^\circ = 60' \text{ (minutes longitude)}$$

$$1' = 60'' \text{ (seconds longitude)}$$

Sidereal time is defined in the following figure

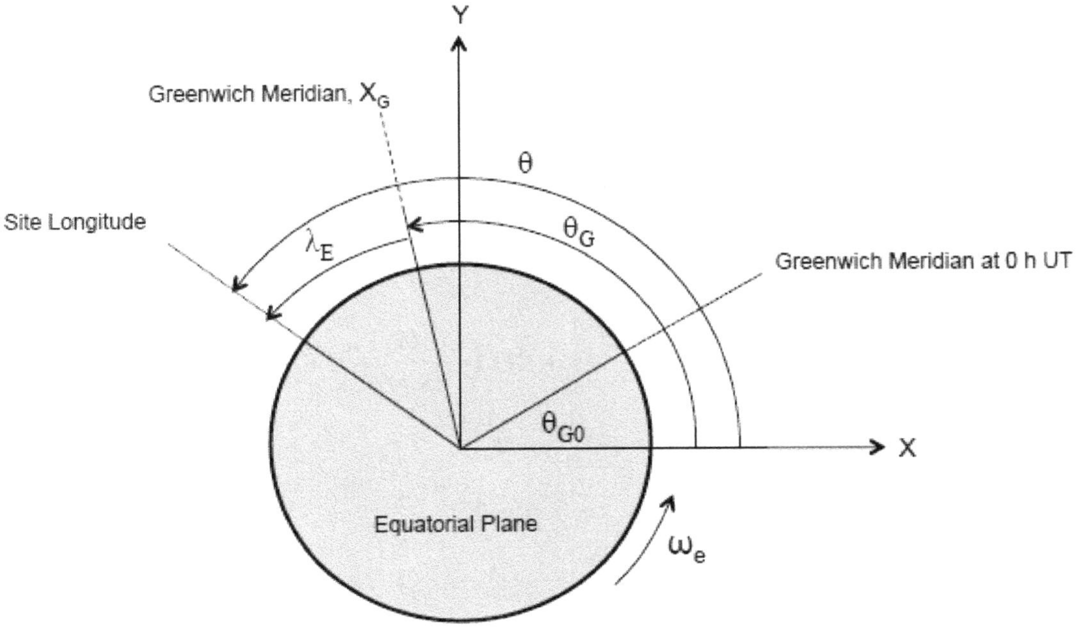


where

$X$ = vernal equinox direction
$\omega_e$ = angular velocity of Earth
$\theta_{G0}$ = Greenwich sidereal time at $0^h$ UT
$\theta_G$ = Greenwich sidereal time
$\lambda_E$ = east longitude of the site, i.e., the angle measured eastward in the equatorial plane between the Greenwich Meridian at the site's meridian
$\theta$ = sidereal time at the site
$t - t_0$ = the time <u>in minutes after midnight</u> on the date of interest

The sidereal time at the site at a specific JD can be calculated with the following equations

$$T_0 = \frac{J_0 - 2{,}451{,}545.0}{36{,}525} = \text{Julian Century} \quad \text{(based on the J2000 reference)}$$

$$\theta_{G_0} = 100.4606184 + (36{,}000.77004)T_0 + (0.000387933)T_0^2 - (2.583\text{x}10^{-3})T_0^3 \; ^\circ$$

$$\theta_G = \theta_{G_0} + \frac{UT}{24} = \theta_{G_0} + (t - t_0)\omega_e$$

$$\theta = \theta_G + \lambda_E \quad (\text{always adjust } \theta_{G_0}, \theta_G, \text{and } \theta \text{ to be} \leq 360^\circ)$$

Example 2.2

Calculate the local sidereal time at the Greenwich Meridian at a UT of 12:45:00 AM on February 12, 2014.

Solution:

$$y = 2014 \quad , \quad M = 2 \quad , \quad d = 12 \quad , \quad h = m = s = 0 \quad \text{at } 0^h \text{ UT}$$

$$t - t_0 = 45.0 \text{ min}$$

$$J_0 = 367y - \text{INT}\left\{\frac{7\left[y + \text{INT}\left(\frac{M+9}{12}\right)\right]}{4}\right\} + \text{INT}\left(\frac{275M}{9}\right) + d + 1{,}721{,}013.5 \text{ days}$$

$$J_0 = 367(2014) - \text{INT}\left\{\frac{7(2014) + \text{INT}\left(\frac{2+9}{12}\right)}{4}\right\} + \text{INT}\left\{\frac{275(2)}{9}\right\} + 12 + 1{,}721{,}013.5$$

$$J_0 = 739{,}138 - 3{,}524 + 61 + 12 + 1{,}721{,}013.5 = 2{,}456{,}700.5 \text{ days}$$

$$T_0 = \frac{J_0 - 2{,}451{,}545.}{36{,}525} = \frac{2{,}456{,}700.5 - 2{,}451{,}545.}{36{,}525} = 0.141150 \text{ cent}$$

$$\theta_{G_0} = 100.4606184 + (36{,}000.77004)T_0 + (0.000387933)T_0^2 - (2.583\text{x}10^{-3})T_0^3$$

$$\theta_{G_0} = 100.4606184 + (36{,}000.77004)(0.141150) + (0.000387933)(0.141150)^2$$

$$-(2.583\text{x}10^{-3})(0.141150)^3 = 5{,}181.965620^\circ \equiv 141.965620^\circ$$

$$\theta_G = \theta_{G_0} + (t - t_0)\omega_e = 141.965620^\circ + (45 \text{ min})(2.507 \times 10^{-1} \tfrac{^\circ}{\text{min}})$$

$$\Rightarrow \theta_G = 153.247120^\circ$$

Example 2.3

Compute the local sidereal time in Auburn, AL at 1:30:45 PM on January 24, 2013.

Solution:

$$\lambda_E = 274.516873^o \quad , \quad UT = 13:30:45 + 6.0 = 19:30:45$$

$$y = 2013 \quad , \quad M = 1 \quad , \quad d = 24 \quad , \quad h = 19 \quad , \quad m = 30 \quad , \quad s = 45$$

$$J_0 = 367y - INT\left\{\frac{7\left[y + INT\left(\frac{M+9}{12}\right)\right]}{4}\right\} + INT\left(\frac{275M}{9}\right) + d + 1{,}721{,}013.5 \text{ days}$$

$$J_0 = 367(2013) - INT\left\{\frac{7(2013)+INT\left(\frac{1+9}{12}\right)}{4}\right\} + INT\left\{\frac{275(1)}{9}\right\} + 24 + 1{,}721{,}013.5$$

$$J_0 = 738{,}771 - 3{,}522 + 30 + 24 + 1{,}721{,}013.5 = 2{,}456{,}316.5 \text{ days}$$

$$T_0 = \frac{J_0 - 2{,}451{,}545.}{36{,}525} = \frac{2{,}456{,}316.5 - 2{,}451{,}545.0}{36{,}525} = 0.13063655 \text{ cent}$$

$$\theta_{G_0} = 100.4606184 + (36{,}000.77004)T_0 + (0.000387933)T_0^2 - (2.583x10^{-3})T_0^3$$

$$\Theta_{G_0} = 100.4606184 + (36{,}000.77004)(0.13063655)$$

$$+ (0.000387933)(0.13063655)^2 - (2.583x10^{-3})(0.13063655)$$

$$\theta_{G_0} = 4{,}803.477031^o \equiv 123.477031^o$$

$$t - t_0 = 19(60) + 30 + \frac{45}{60} = 1{,}170.5 \text{ min}$$

$$\theta_G = \theta_{G_0} + (t - t_0)\omega_e = 123.477031^o + (1{,}170.5)(2.507 \times 10^{-1})$$

$$\theta_G = 416.965874^o - 360^o = 56.965874^o$$

$$\theta = \theta_G + \lambda_E = 56.965874^o + 274.516873^o$$

$$\Rightarrow \theta = 331.482747^o$$

Problems

2.1 Calculate the Julian Date for the times below.
(a) 15:00 UT on February 2, 1952
(b) 12:00 UT on April 1, 2013
(Ans. (a) JD = 2,434,045.125 days; (b) JD = 2,456,384.0 days)

2.2 Find the Julian Date at the times given.
(a) 6:44:56.3 UT on April 12, 1980
(b) 15:30:19.4 UT on May 29, 1910
(c) 1:23:39.35 UT on August 19, 1951
(Ans. (a) JD = 2,444,341.781 days; (b) JD = 2,418,821.146058 days;
(c) JD = 2.433,877.558095 days)

2.3 Compute the number of days between the start of the day in Greenwich, England on March 8, 1921, and the start of the day on August 19, 1983.
(Ans. $\Delta t = 22{,}809$ days)

2.4 Determine the Julian Date on May 5, 2022, at a time of 11:45 AM in New Orleans, LA.
(Ans. JD = 2,459,705.23958333 days)

2.5 Calculate the local sidereal time at the Greenwich Meridian at a UT of 11:15:00 PM on March 11, 2012.
(Ans. $\theta_G = 158.782062°$)

2.6 Compute the local sidereal time at Belgrade, Serbia, where $\phi = 44°52'$, $\lambda_E = 20°32'$ at 12:15 UT on March 31, 2009.
(Ans. $\theta = 33.2878515°$)

2.7 The local sidereal time at an unknown site in Texas is $\theta = 193.326826°$ at 7:34:21 PM on July 2, 1987. Determine the west longitude of the site.
(Ans. $\lambda_W = 97.743061°$)

2.8 In Rio de Janeiro, Brazil, $\lambda_W = 43°06'$, compute the local sidereal time at 3:00 UT on February 15, 2006.
(Ans. $\theta = 146.8840556°$)

# Module 3: Newton's Laws of Motion and Universal Gravitation

The two-body problem describes the motion of two point masses in space affected only by their mutual gravitational attraction. It's very useful in understanding orbital mechanics but it rarely, if ever, exists in nature. The two-body problem is idealistic because two major and unrealistic assumptions must be made in order to develop the equations of motion of the two bodies and solve the problem. Those assumptions are

- Both bodies are considered to be point masses, i.e., their masses are concentrated at a single point in space.

- The mutual gravitational attraction of the masses is the only force acting on the bodies. No other forces affect their motion.

Neither of those assumptions is a good one in the real world so use of the two-body equations of motion has limitations. However, much can be learned from studying their use and they serve as the baseline for actual problems where the two assumptions above are inaccurate.

The two-body problem is a difficult one and the only problem that can be analytically solved. The solution is obtained through reliance on the two broad assumptions described above. Real world problems where one or both masses are not point masses and the existence of other forces affecting the motion of each body, i.e., perturbation forces, are much more difficult. No closed-form analytical solution exists for any problem more complicated than the two-body problem. The two-body equations of motion will be developed using Newton's Laws of Motion and his Law of Universal Gravitation.

## Newton's Laws of Motion

Isaac Newton published his three Laws of Motion and his Law of Universal Gravitation in 1687 in *Philosophiae Naturalis Principia Mathematica* (*The Mathematical Principles of Natural Philosophy*), widely known as *Principia*.

The development of Newton's Laws is considered to be one of the supreme achievements of the human mind. This work was completed over 20 years previously but Newton was hesitant to publish it due to his adverse fear of criticism. He only published this groundbreaking work at the insistence of his colleagues.

The first fact to understand about Newton's Laws is that they're only valid in an inertial coordinate system.

An inertial coordinate system one that is either at rest, or moving in a straight line with constant speed, i.e., rectilinear motion. In other words, they are only valid in non-accelerating coordinate systems.

Any coordinate system that moves on a circular path, even if it has constant speed, has a centripetal, i.e., normal, acceleration directed axially inward on that path.

The origin of an inertial coordinate system must be located at an 'inertial point', which also possesses the property of being at rest or moving with rectilinear motion.

Newton's First Law can be stated as: *A body continues in its state of rest, or uniform motion in a straight line, unless acted upon by a force.* This law is known as the 'equilibrium' law.

Newton's Second Law can be stated as: *The rate-of-change of linear momentum is proportional to the impressed force and takes place in the direction in which the force acts.*

Newton's Second Law is known as the 'F = ma' law, but that isn't exactly accurate. This law can be written

$$\bar{F} = \frac{d(m\bar{v})}{dt}$$

$$\bar{F} = \left(\frac{dm}{dt}\right)\bar{v} + m\left(\frac{d\bar{v}}{dt}\right)$$

$$\Rightarrow \bar{F} = \dot{m}\bar{v} + m\bar{a}$$

If the mass is constant, i.e., $\dot{m} = 0$, then $\bar{F} = m\bar{a}$. However, if the mass is changing, as in the case of a thrusting rocket or spacecraft, then $\bar{F} = m\bar{a}$ is <u>not</u> applicable.

Newton's Third Law can be stated as: *For every action there is always an equal and opposite reaction.* This is known as the 'action/reaction' law, which can be expressed

$$\bar{F}_{12} = \bar{F}_{21}$$

## Newton's Law of Universal Gravitation

Newton's Law of Universal Gravitation can be stated as: *Any two particles attract each other with a force directed along the line adjoining them, and which has a magnitude directly proportional to the product of the masses and inversely proportional to the square of the distance between them.*

The magnitude of this gravitational force is written

$$F = \frac{Gm_1m_2}{r^2}$$

where

$G =$ Constant of Universal Gravitation
$\bar{r} =$ position vector of $m_2$ relative to $m_1$
$r = |\bar{r}| =$ the distance between $m_1$ and $m_2$

The gravitational force vector can be expressed

$$\bar{F} = \frac{Gm_1m_2}{r^3}\bar{r}$$

Isaac Newton, through the development of Newton's Laws, was the first to provide an explanation of the dynamical motion of bodies in space, and not just a description.

### Example 3.1

Determine the distance from the center of Saturn to a point between Saturn and the Sun where the magnitudes of the gravitational forces due to both bodies are equal.

Solution:

$$F_{Sun} = F_{Sat}$$

Let x equal the distance from the center of Saturn to the point where the forces from each body are equal.

$$\frac{Gm_{Sun}m}{(a_{Sat}-x)^2} = \frac{Gm_{Sat}m}{x^2}$$

$$m_{Sun}x^2 = m_{Sat}(a_{Sat} - x)^2$$

$$m_{Sun}x^2 = m_{Sat}(a_{Sat}^2 - 2a_{Sat}x + x^2)$$

$$(m_{Sun} - m_{Sat})x^2 + 2m_{Sat}a_{Sat}x - m_{Sat}a_{Sat}^2 = 0$$

$$(1.989 \times 10^{30} - 5.685 \times 10^{26})x^2 + 2(5.685 \times 10^{26})(1.433 \times 10^9)x$$

$$-(5.685 \times 10^{26})(1.433 \times 10^9) = 0$$

$$x^2 + (8.198 \times 10^5)x - 5.869 \times 10^{14} = 0$$

$$\Rightarrow \ x = 2.382 \times 10^7 \text{ km}$$

Example 3.2

Calculate the magnitude of the gravitational acceleration acting on a satellite at an altitude of 300 km above the surface of Mars.

Solution

$$r = R_M + 300 = 3{,}396 + 300 = 3{,}696 \text{ km}$$

$$a = \frac{F}{m} = \frac{\frac{Gm_M m}{r^2}}{m} = \frac{Gm_M}{r^2} = \frac{(6.6742 \times 10^{-20})(6.419 \times 10^{23})}{(3{,}696)^2}$$

$$\Rightarrow \ a = 3.136 \times 10^{-3} \ \frac{\text{km}}{\text{s}^2}$$

Example 3.3

Find the weight of a person standing on the surface of Jupiter if they weigh 200 lbs on the surface of Earth.

Solution:

$$F_J = \frac{GmM_J}{R_J^2} = \frac{Gm(1.899 \times 10^{27})}{(71{,}490)^2} = 3.716 \times 10^{17} Gm$$

$$F_E = \frac{GmM_E}{R_E^2} = \frac{Gm(5.974 \times 10^{24})}{(6{,}378.1)^2} = 1.469 \times 10^{17} Gm$$

$$W_J = \left(\frac{F_J}{F_E}\right) W_E = \left(\frac{3.716 \times 10^{17} Gm}{1.469 \times 10^{17} Gm}\right)(200) = \left(\frac{3.716 \times 10^{17}}{1.469 \times 10^{17}}\right)(200)$$

$$\Rightarrow \ W_J = 505.9 \text{ lbs}$$

Problems

3.1 Determine the distance from the center of Jupiter to a point between Jupiter and the Sun where the magnitudes of the gravitational forces from both bodies are equal.
(Ans. $2.334 \times 10^7$ km from the center of Jupiter)

3.2 Calculate the weight of a person standing on the surface of the moon if they weigh 200 lbs on the surface of Earth.
(Ans. 33.2 lbs)

3.3 Compute the weight of a person standing on the surface of Mars if they weigh 200 lbs on the surface of Earth.
(Ans. 75.69 lbs)

3.4 Perform an analysis of the force between the Sun and moon and Earth and moon to determine if the moon orbits the Sun or Earth.
(Ans. $\frac{F_S}{F_E} = 2.198 \implies$ the moon orbits the Sun)

3.5 Calculate the approximate value of the acceleration of gravity on Earth's surface.
(Ans. $\bar{a} = 9.80 \frac{km}{s^2}$)

3.6 Determine the distance from Earth's center to a point between Earth and the moon where the magnitudes of the gravitational forces from Earth and the moon are equal.
(Ans. 346,036.0 km from the center of Earth)

3.7 Calculate the gravitational acceleration vector (in the ECI coordinate system) of a satellite located at an altitude of 1,000 km above the North Pole.
(Ans. $\bar{a} = -7.322 \times 10^{-3}\ \bar{K}\ \frac{km}{s^2}$)

3.8 Find the mass of a celestial body located $2.5 \times 10^7$ km from Earth if the point between them where the gravitational forces from both bodies are equal is located at a distance of $2.0 \times 10^7$ km from Earth.
(Ans. $m = 3.734 \times 10^{23}$ kg)

(This page was intentionally left blank.)

# Module 4: Two-Body Inertial Equations of Motion

The development of the inertial equations of motion for the two-body problem begins with a single point mass, $m_1$, located by position vector, $\overline{R}_1$ in an inertial coordinate system centered at inertial point O.

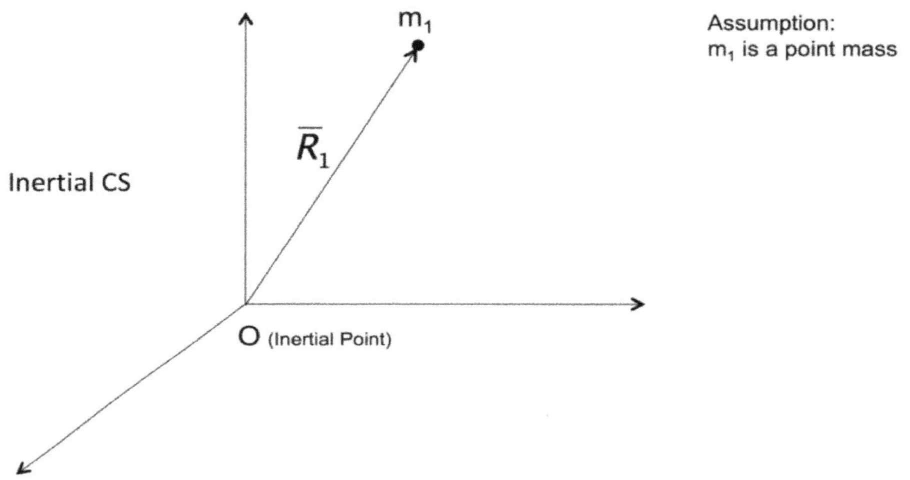


Newton's First Law states that mass $m_1$ will be in the state of equilibrium, i.e., at rest or in uniform motion, unless a force acts upon it to change its state.

But what forces act on $m_1$?

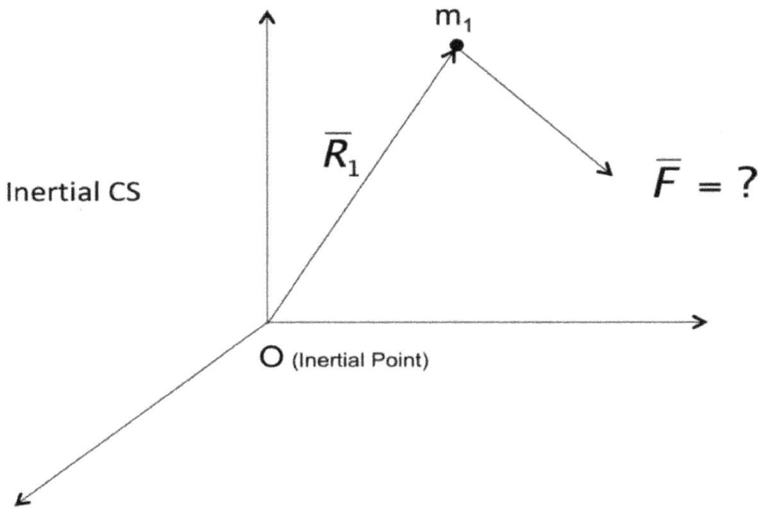

Consider the addition of a second point mass, $m_2$, located by position vector, $\overline{R}_2$ in the inertial coordinate system as shown below.

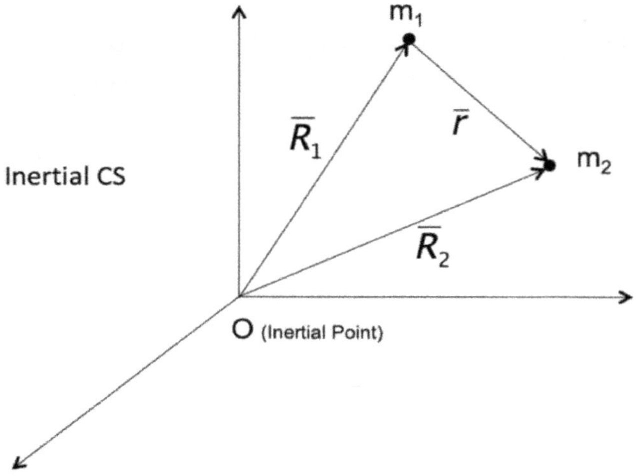


Assumption:
$m_1$ and $m_2$ are point masses

where

$\bar{r} =$ the position vector of $m_2$ <u>relative</u> to $m_1$

Based on Newton's Law of Universal Gravitation, the two masses undergo a mutual gravitational attraction. Newton's Third Law states that these forces must be equal in magnitude and opposite in direction as

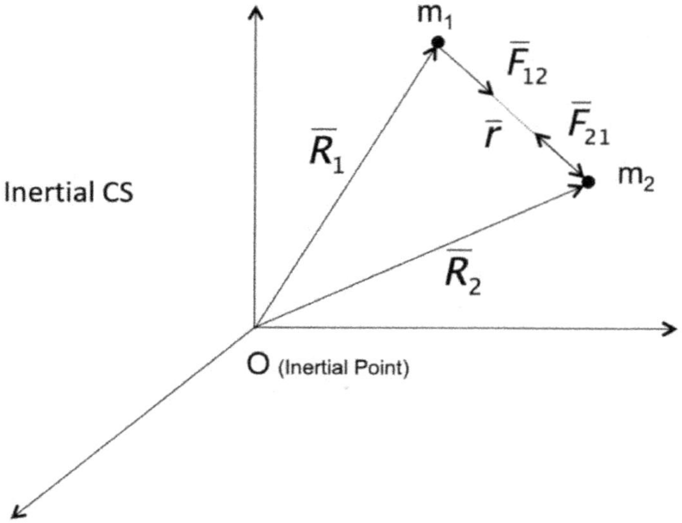


$$\overline{F}_{21} = -\overline{F}_{12}$$

Review of assumptions of the two-body problem:

- The coordinate system is inertial, with an origin located at an 'inertial point'.
- Both masses are considered to be point masses.
- The only force present is the mutual gravitational attraction of the two masses.

Under these assumptions, no other forces will affect the motion of the masses. With the application of Newton's Second Law and the Law of Universal Gravitation, the equations of motion in the inertial coordinate system for each mass can be written

$$m_1\ddot{\bar{R}}_1 = \frac{Gm_1m_2}{r^3}\bar{r}$$

$$m_2\ddot{\bar{R}}_2 = -\frac{Gm_1m_2}{r^3}\bar{r} \qquad (1)$$

Each equation above is a second order, three-dimensional vector equation and represents a $6^{th}$ order (one equation x $2^{nd}$ order x three dimensions) ordinary differential equation.

Since the equations are coupled through the term r, Equations (1) represent a $12^{th}$ order system of ordinary differential equations.

To solve the $12^{th}$ order system, twelve constants of integration (or <u>integrals</u>) are required. Integration of Equations (1) twice will provide the positions and velocities of both masses in the inertial coordinate system. However, Equations (1) can't be integrated directly because the right sides of both equations are nonlinear.

Therefore, additional analytical relationships must be found to provide sufficient information to obtain solutions to the system of equations. This information will correspond to the twelve constants of integration needed.

As a scalar example of integrals, consider the relationship

$$a(t) = 0$$

Integrating directly once gives

$v(t) = v_0$ ($v_0$ being known gives one integral)

Integrating directly again gives

$s(t) = v_0(t) + s_0$ ($s_0$ being known gives a second integral)

The initial conditions, $v_0$ and $s_0$, together represent two total integrals and the position and velocity as functions of time can be obtained.

For a three-dimensional vector example of integrals, consider the relationship

$$\bar{a}(t) = 0$$

Integrating directly once gives

$$\bar{v}(t) = \bar{v}_0 \qquad (\bar{v}_0 \text{ being known gives three integrals})$$

Integrating directly again gives

$$\bar{r}(t) = \bar{v}_0(t) + \bar{r}_0 \qquad (\bar{r}_0 \text{ being known gives three more integrals})$$

The initial conditions, $\bar{v}_0$ and $\bar{r}_0$, together represent six total 'integrals' and the position and velocity vectors as functions of time can be obtained.

<u>Center-of-Mass Integrals</u>

Adding Equations (1) together gives

$$m_1\ddot{\bar{R}}_1 + m_2\ddot{\bar{R}}_2 = 0$$

Direct integration shows

$$m_1\dot{\bar{R}}_1 + m_2\dot{\bar{R}}_2 = \bar{A} \qquad (\bar{A} \text{ being known gives three integrals})$$

Direct integration again yields

$$m_1\bar{R}_1 + m_2\bar{R}_2 = \bar{A}t + \bar{B} \qquad (\bar{B} \text{ being known gives three additional integrals})$$

Dividing both of the equations above by the quantity $(m_1 + m_2)$ gives

$$\frac{m_1\dot{\bar{R}}_1+m_2\dot{\bar{R}}_2}{m_1+m_2} = \frac{\bar{A}}{m_1+m_2}$$

$$\frac{m_1\bar{R}_1+m_2\bar{R}_2}{m_1+m_2} = \frac{\bar{A}t+\bar{B}}{m_1+m_2}$$

The position vector to the center-of-mass, $\bar{R}_{CM}$, is shown on the following figure.

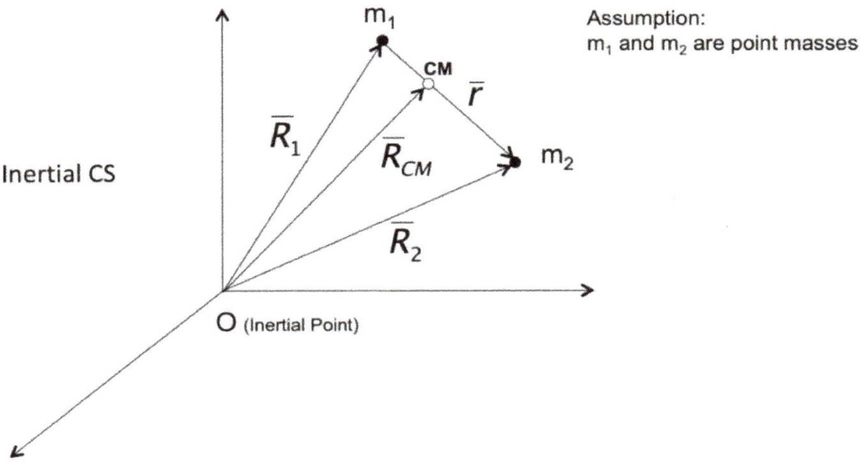


By definition, the position vector of the center-of-mass is

$$\bar{R}_{CM} = \frac{m_1\bar{R}_1+m_2\bar{R}_2}{m_1+m_2} = \frac{\bar{A}t+\bar{B}}{m_1+m_2} = \text{linear function of time}$$

The time derivative of position vector gives the velocity vector of the center-of-mass as

$$\dot{\bar{R}}_{CM} = \frac{m_1\dot{\bar{R}}_1+m_2\dot{\bar{R}}_2}{m_1+m_2} = \frac{\bar{A}}{m_1+m_2} = \text{constant}$$

These results show that the position of the center-of-mass of the two-body system changes linearly with time, i.e., it has rectilinear motion, and its velocity is constant.

Therefore, the center-of-mass of the system has the properties of an inertial point and can be used as the origin of an inertial coordinate system.

The inertial coordinate system discussed above can then be re-defined with its origin at the center-of-mass of the two-body system, as

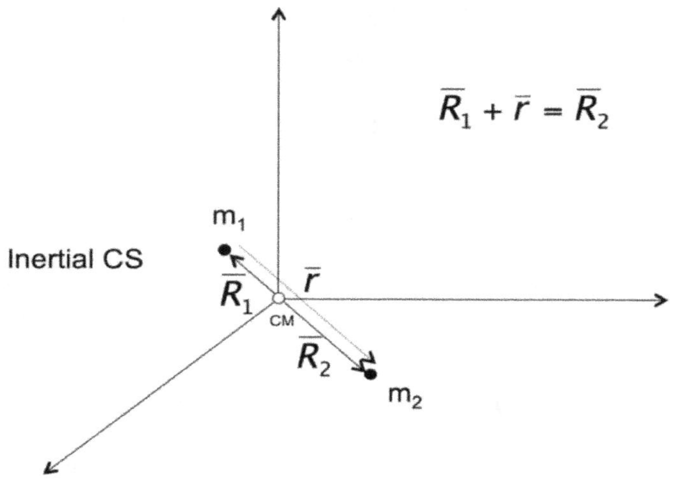


This indicates that both masses will orbit about the center-of-mass and always remain directly opposite each other as shown below.

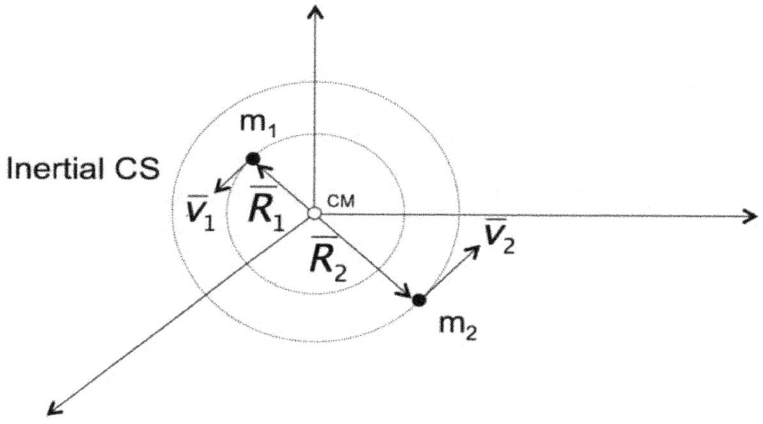


In the re-defined inertial coordinate system, the position vector from the origin to the center-of-mass is zero. Therefore, the position and velocity of the center-of-mass simplify to

$$\dot{\bar{R}}_{CM} = \frac{\bar{A}t+\bar{B}}{m_1+m_2} = 0$$

$$\bar{R}_{CM} = \frac{\bar{A}}{m_1+m_2} = 0$$

These equations can only be satisfied or <u>all</u> values of time if $\bar{A} = \bar{B} = 0$.

Therefore, it follows that

$$m_1\dot{\bar{R}}_1 + m_2\dot{\bar{R}}_2 = 0$$
$$m_1\bar{R}_1 + m_2\bar{R}_2 = 0 \qquad (2)$$

These two vector equations, or six scalar equations, are known as the Center-of-Mass Integrals and represent six integrals of the motion of the two-body problem in an inertial coordinate system whose origin is at the center-of-mass of the system.

<u>Example 4.1</u>

In a two-dimensional ECI coordinate system mass $m_1$ has position and velocity vectors of $\bar{R}_1 = 10{,}000\ \bar{I}$ km and $\bar{V}_1 = -4\ \bar{J}\ \frac{km}{s}$. Calculate the position and velocity vectors of a second mass $m_2$, if $m_2 = 2m_1$.

Solution:

$$m_1\bar{R}_1 + m_2\bar{R}_2 = 0$$

$$\bar{R}_2 = -\left(\frac{m_1}{m_2}\right)\bar{R}_1 = -\left(\frac{m_1}{2m_1}\right)(10{,}000\ \bar{I})$$

$$\Rightarrow\ \bar{R}_2 = -5{,}000\ \bar{I}\ \text{km}$$

$$m_1\dot{\bar{R}}_1 + m_2\dot{\bar{R}}_2 = 0$$

$$\dot{\bar{R}}_2 = -\left(\frac{m_1}{m_2}\right)\dot{\bar{R}}_1 = -\left(\frac{m_1}{2m_1}\right)(-4\ \bar{J})$$

$$\Rightarrow\ \dot{\bar{R}}_2 = 2.0\ \bar{J}\ \frac{km}{s} = \bar{V}_2$$

Example 4.2

For a two point-mass system, use the data below to find $\overline{R}_1$ and $\overline{V}_2$ in the ECI coordinate system.

$$m_1 = 5.0 \times 10^5 \text{ kg}$$
$$m_2 = 8.0 \times 10^5 \text{ kg}$$
$$\overline{V}_1 = 3.0\,\bar{I} - 4.0\,\bar{J} - 2.5\,\overline{K}\frac{\text{km}}{\text{s}}$$
$$\overline{R}_2 = 3{,}000\,\bar{I} + 7{,}500\,\bar{J} - 2{,}800\,\overline{K}\text{ km}$$

Solution:

$$m_1\overline{R}_1 + m_2\overline{R}_2 = 0$$

$$\overline{R}_1 = -\left(\frac{m_2}{m_1}\right)\overline{R}_2 = -\left(\frac{8\times10^5}{5\times10^5}\right)(3{,}000\,\bar{I} + 7{,}500\,\bar{J} - 2{,}800\,\overline{K})$$

$$\Rightarrow \overline{R}_1 = -4{,}800\,\bar{I} - 12{,}000\,\bar{J} + 4{,}480\,\overline{K}\ \text{ km}$$

$$m_1\dot{\overline{R}}_1 + m_2\dot{\overline{R}}_2 = 0$$

$$\dot{\overline{R}}_2 = -\left(\frac{m_1}{m_2}\right)\overline{V}_1 = -\left(\frac{5\times10^5}{8\times10^5}\right)(3.0\,\bar{I} - 4.0\,\bar{J} - 2.5\,\overline{K}) = \overline{V}_2$$

$$\Rightarrow \overline{V}_2 = -1.88\,\bar{I} + 2.50\,\bar{J} + 1.56\,\overline{K}\ \frac{\text{km}}{\text{s}}$$

Since twelve integrals are needed to solve the Two-Body Inertial Equations of Motion, and the Center-of-Mass Integrals provide six, six more integrals are required to obtain a solution to the Two-Body Problem in inertial coordinates.

The Center-of-Mass Integrals provide

- Six of the twelve integrals needed for a solution.

- The property that the sum of each mass times its position vector must always equal zero.

- The property that the sum of each mass times its velocity vector must always equal zero.

Consider the relationships

$$\dot{\bar{R}}_1 + \dot{\bar{r}} = \dot{\bar{R}}_2$$

$$\bar{R}_1 + \bar{r} = \bar{R}_2$$

Substitution of the above expressions into Equations (2) yields the results

$$\dot{\bar{R}}_1 = \frac{-m_2}{m_1+m_2}\dot{\bar{r}}$$

$$\bar{R}_1 = \frac{-m_2}{m_1+m_2}\bar{r}$$

These equations give the solution for the position and velocity of mass $m_1$ in terms of the position and velocity of mass $m_2$ relative to mass $m_1$.

Example 4.3

Consider two point masses, $m_1 = 9.0 \times 10^{24}$ kg and $m_2 = 2.0 \times 10^{24}$ kg. In an inertial coordinate system, the position and velocity of $m_2$ are

$$\bar{R}_2 = (3.0\,\bar{I} + 5.2\,\bar{J} - 2.4\,\bar{K}) \times 10^5 \text{ km} \qquad \bar{V}_2 = 1.0\,\bar{I} - 2.0\bar{J} + 1.75\,\bar{K}\,\frac{\text{km}}{\text{s}}$$

Determine the position and velocity of $m_2$ relative to $m_1$.

Solution:

$$m_1\bar{R}_1 + m_2\bar{R}_2 = 0$$

$$\bar{R}_1 = \left(\frac{-m_2}{m_1}\right)\bar{R}_2 = \left(\frac{-2.0\times10^{24}}{9.0\times10^{24}}\right)(3.0\,\bar{I} + 5.2\,\bar{J} - 2.4\,\bar{K}) \times 10^5$$

$$\bar{R}_1 = (-0.67\,\bar{I} - 1.16\,\bar{J} + 0.53\,\bar{K}) \times 10^5 \text{ km}$$

$$\bar{R}_1 = \frac{-m_2}{m_1+m_2}\bar{r}$$

$$\bar{r} = \frac{-(m_1+m_2)}{m_2}\bar{R}_1 = \frac{-(9.0\times10^{24}+2.0\times10^{24})}{2.0\times10^{24}}(-0.67\,\bar{I} - 1.16\,\bar{J} + 0.53\,\bar{K}) \times 10^5$$

$$\Rightarrow \bar{r} = (3.67\bar{I} + 6.36\,\bar{J} - 2.93\,\bar{K}) \times 10^5 \text{ km}$$

$$m_1\dot{\bar{R}}_1 + m_2\dot{\bar{R}}_2 = 0$$

$$\dot{\bar{R}}_1 = \left(\frac{-m_2}{m_1}\right)\dot{\bar{R}}_2 = \left(\frac{-2.0\times10^{24}}{9.0\times10^{24}}\right)(1.0\,\bar{I} - 2.0\,\bar{J} + 1.75\,\bar{K})$$

$$\dot{\bar{R}}_1 = -0.22\,\bar{I} + 0.44\,\bar{J} - 0.39\,\bar{K}\,\frac{km}{s}$$

$$\dot{\bar{R}}_1 = \frac{-m_2}{m_1+m_2}\dot{\bar{r}}$$

$$\dot{\bar{r}} = \frac{-(m_1+m_2)}{m_2}\dot{\bar{R}}_1 = \frac{-(9.0\times10^{24}+2.0\times10^{24})}{2.0\times10^{24}}(-0.22\,\bar{I} + 0.44\bar{J} - 0.39\,\bar{K})$$

$$\Rightarrow\ \dot{\bar{r}} = 1.22\,\bar{I} - 2.44\,\bar{J} + 2.14\,\bar{K}\,\frac{km}{s}$$

Problems

4.1 A new solar system has been discovered consisting of a Sun and two planets. The mass of the new Sun has been determined to be $4.2 \times 10^{30}$ kg. The center-of-mass for the Sun-Planet 1 system is a distance of 2,678,800 km from the Sun, with Planet 1 being a distance of $8.540 \times 10^{10}$ km from the Sun. Calculate the mass of Planet 1.
(Ans. $m_{P1} = 1.317 \times 10^{26}$ kg)

4.2 In the solar system described in Problem 4.1, if Planet 2 has a mass of $15.1 \times 10^{27}$ kg and the center-of-mass for the Sun-Planet 2 system is a distance of 475,075,000.0 km from the Sun. Determine the distance of Planet 2 from the center-of-mass.
(Ans. $1.321 \times 10^{11}$ km)

4.3 Derive the inertial equations of motion for each body in a system of three point masses moving under their mutual gravitational attraction like Equations (1) provided for a two point mass system.
(Use $\bar{r}_{12}$ = position vector of $m_2$ relative to $m_1$
$\bar{r}_{13}$ position vector of $m_3$ relative to $m_1$, and
$\bar{r}_{23}$ = position vector of $m_3$relative to $m_2$)
(Ans. $m_1\ddot{\bar{R}}_1 = \frac{Gm_1m_2}{r_{12}^3}\bar{r}_{12} + \frac{Gm_1m_3}{r_{13}^3}\bar{r}_{13}$, $m_2\ddot{\bar{R}}_2 = -\frac{Gm_1m_2}{r_{12}^3}\bar{r}_{12} + \frac{Gm_2m_3}{r_{23}^3}\bar{r}_{23}$,
$m_3\ddot{\bar{R}}_3 = -\frac{Gm_1m_3}{r_{13}^3}\bar{r}_{13} - \frac{Gm_2m_3}{r_{23}^3}\bar{r}_{23}$)

4.4 In a solar system consisting of a Sun and two planets the mass of the Sun is $3.7 \times 10^{30}$ kg. If the center-of-mass for the Sun-Planet 1 system is a distance of 2,678,800.0 km from the Sun and Planet 1 is a distance of $8.540 \times 10^{10}$ km from the Sun, calculate the mass of Planet 1.
(Ans. $m_{P1} = 1.16 \times 10^{26}$ kg)

4.5 Consider the system of masses described in Problem 4.4. If Planet 2 has a mass of $16.2 \times 10^{27}$ kg and the center-of-mass for the Sun-Planet 2 system is a distance of 475,049,600 km from the Sun, find the distance of Planet 2 from the center-of-mass.
(Ans. $1.085 \times 10^{11}$ km)

4.6 Consider two point masses $m_1 = 8.0 \times 10^{24}$ kg, $m_2 = 3.0 \times 10^{24}$kg. In an inertial coordinate system, the position and velocity of mass $m_2$ are

$$\bar{R}_2 = (2.5\,\bar{I} + 5.0\,\bar{J} - 3.0\,\bar{K}) \times 10^5 \text{ km} \qquad \bar{V}_2 = 1.1\,\bar{I} - 2.1\,\bar{J} + 1.7\,\bar{K}\,\frac{\text{km}}{\text{s}}$$

Determine the position and velocity of $m_2$ <u>relative</u> to mass $m_1$,
(Ans. $\bar{r} = (3.44\,\bar{I} + 6.88\,\bar{J} - 4.13\,\bar{K}) \times 10^5$ km, $\bar{v} = 1.51\,\bar{I} - 2.89\,\bar{J} + 2.34\,\bar{K}\,\frac{\text{km}}{\text{s}}$)

4.7 For the masses given in Problem 4.6, determine the acceleration of both masses $m_1$ and $m_2$ in the inertial coordinate system.
(Ans. $\ddot{\bar{R}}_1 = (1.04\,\bar{I} + 2.07\,\bar{J} - 1.24\,\bar{K}) \times 10^{-7}\,\frac{\text{km}}{\text{s}^2}$),
$\ddot{\bar{R}}_2 = (-2.76\,\bar{I} - 5.53\,\bar{J} + 3.32\,\bar{K}) \times 10^{-7}\,\frac{\text{km}}{\text{s}^2}$)

4.8 Using the results of Problem 4.7, find the acceleration of mass $m_1$ <u>relative</u> to mass $m_2$.
(Ans. $\ddot{\bar{r}} = (3.80\,\bar{I} + 7.60\,\bar{J} - 4.56\,\bar{K}) \times 10^{-7}\,\frac{\text{km}}{\text{s}^2}$)

(This page was intentionally left blank.)

# Module 5: Two-Body Relative Equations of Motion

It's generally more useful and more convenient to determine the motion of one body relative to another body, rather than their motion relative to the center-of-mass of a system of masses, e.g., the motion of a satellite relative to Earth.

The Center-of-Mass Integrals make it possible to view the two-body problem as a 'relative motion' problem rather than an inertial motion problem.

Since

$$\bar{R}_1 + \bar{r} = \bar{R}_2$$

Differentiation twice shows

$$\ddot{\bar{R}}_1 + \ddot{\bar{r}} = \ddot{\bar{R}}_2$$

$$\Rightarrow \ddot{\bar{r}} = \ddot{\bar{R}}_2 - \ddot{\bar{R}}_1$$

Substituting Equations (1) into this expression gives

$$\ddot{\bar{r}} = \frac{Gm_1}{r^3}\bar{r} - \frac{Gm_2}{r^3}\bar{r}$$

$$\ddot{\bar{r}} = \frac{-G(m_1 + m_2)}{r^3}\bar{r}$$

This equation is commonly written as

$$\ddot{\bar{r}} = \frac{-\mu}{r^3}\bar{r} \tag{3}$$

where

$\mu = G(m_1 + m_2) =$ gravitational parameter for masses $m_1$ and $m_2$

Equation (3) is known as the 'Two-Body Relative Equations of Motion'. They represent the acceleration of mass $m_2$ relative to mass $m_1$.

Integration of Equation (3) twice gives the position and velocity of mass $m_2$ relative to mass $m_1$ (as if mass $m_1$ was stationary).

Converting the equations of motion from the inertial coordinate system to a relative coordinate system results in moving the origin to the center of mass $m_1$. This change provides an origin that is directly observable and the motion of $m_2$ relative to $m_1$ becomes more meaningful. The relative coordinate system is shown as

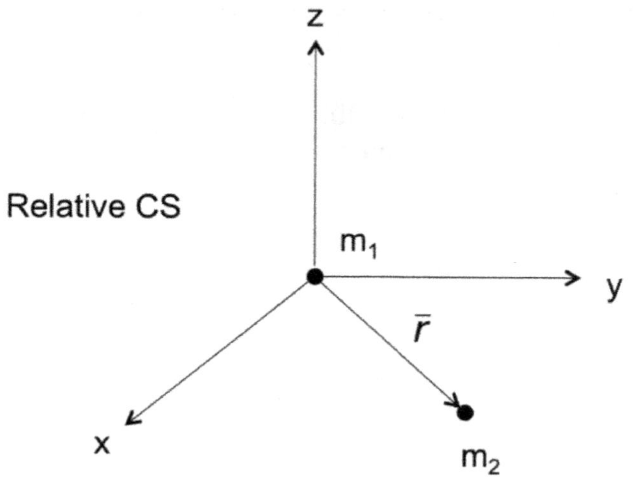


Equations of motion in this relative coordinate system are provided by Equation (3). Solving Equation (3) gives the position and velocity of $m_2$ relative to $m_1$. These relative equations represent a 6th order system (three $2^{nd}$ order vector equations) of ordinary differential equations.

The solution of the $6^{th}$ order system requires six integrals or constants of integration. The conversion of the Two-Body Inertial Equations of Motion to relative equations of motion was made possible by using the six Center-of-Mass Integrals. Doing so reduced the original $12^{th}$ order system of inertial equations to a $6^{th}$ order system of relative equations.

The Two-Body Relative Equations of Motion can be obtained another way besides summing the forces acting on $m_2$.

Define a scalar 'potential function', U, which is the negative of the potential energy, as

$$U = \frac{\mu}{r}$$

where

$$\bar{r} = x\,\bar{\imath} + y\,\bar{\jmath} + z\,\bar{k}$$

$$r = \sqrt{x^2 + y^2 + z^2}$$

Using the gradient function

$$\overline{\nabla} = \frac{\delta}{\delta x}\bar{\imath} + \frac{\delta}{\delta y}\bar{\jmath} + \frac{\delta}{\delta x}\bar{k}$$

It can be shown that

$$\overline{\nabla}U = \frac{-\mu}{r^3}\bar{r}$$

Therefore

$$\ddot{\bar{r}} = \overline{\nabla}U$$

When a force can be written as the gradient of a potential function, that force is a <u>conservative</u> force.

The Two-Body Relative Equations of Motion can also be written in scalar form as

$$\ddot{\bar{r}} = \frac{-\mu}{r^3}\bar{r}$$

$$\ddot{x}\,\bar{\imath} + \ddot{y}\,\bar{\jmath} + \ddot{z}\,\bar{k} = \frac{-\mu(x\,\bar{\imath} + y\,\bar{\jmath} + z\,\bar{k})}{(x^2+y^2+z^2)^{1.5}}$$

Separating this vector equation into scalar components gives

$$\ddot{x} = \frac{-\mu x}{(x^2+y^2+z^2)^{1.5}} = \frac{-\mu x}{r^3}$$

$$\ddot{y} = \frac{-\mu y}{(x^2+y^2+z^2)^{1.5}} = \frac{-\mu y}{r^3} \tag{4}$$

$$\ddot{z} = \frac{-\mu z}{(x^2+y^2+z^2)^{1.5}} = \frac{-\mu z}{r^3}$$

Each scalar equation is a 2$^{nd}$ order ordinary differential equation, which requires two integrals to solve. As a 6$^{th}$ order coupled system, six integrals are required to obtain a solution to Equation (4).

Equations (4) represent a form of the Two-Body Relative Equations of Motion that are suited for a solution by numerical integration.

In solving the Two-Body Relative Equations of Motion, integrals can be developed which provide important information about how one body moves relative to another body. The integrals discussed below are important in studying motion in the Two-Body Problem.

Angular Momentum Integrals

The Two-Body Inertial Equations of Motion represent a $12^{th}$ order system of ordinary differential equations and required twelve integrals to obtain a solution. The relative equations represent a $6^{th}$ order system that requires six integrals for a solution.

Like the inertial equations, the relative equations cannot be integrated directly, so other relationships regarding motion in the two-body system must be found to obtain enough information to provide a viable solution to the problem. In finding these additional relationships, much information about the quality of motion of $m_2$ relative to $m_1$ can be obtained. This information is obtained in the same way Isaac Newton first discovered it.

Beginning with the Two-Body Relative Equations of Motion, take the vector, i.e., cross, product of the position vector with Equation (3)

$$\bar{r} \times \ddot{\bar{r}} = \frac{-\mu}{r^3}(\bar{r} \times \bar{r})$$

This equation can be written in the form

$$\frac{d}{dt}(\bar{r} \times \dot{\bar{r}}) = 0$$

Direct integration gives

$$\bar{r} \, x \, \dot{\bar{r}} = \bar{h} = \text{Constant}$$

Rewriting and defining

$$\bar{r} \times \dot{\bar{r}} = \bar{r} \times \bar{v} = \bar{h} = \text{angular momentum}$$

This shows the angular momentum of $m_2$ relative to $m_1$ is constant.

The Angular Momentum Integrals are then written as

$$\bar{r} \times \bar{v} = \bar{h} = \text{Constant} \qquad (5)$$

Since $\bar{h}$ is a vector, it provides three integrals of motion of the Relative Two-Body Problem. Three more integrals are needed.

In summary, for the motion of $m_2$ about $m_1$, the Angular Momentum Integrals provide the following properties of the motion:

- $\bar{r}$ and $\bar{v}$ define the orbital plane.
- $\bar{h}$ is perpendicular to both $\bar{r}$ and $\bar{v}$.
- $\bar{h}$ is constant in both magnitude and direction.

The angular momentum can be determined in polar coordinates as

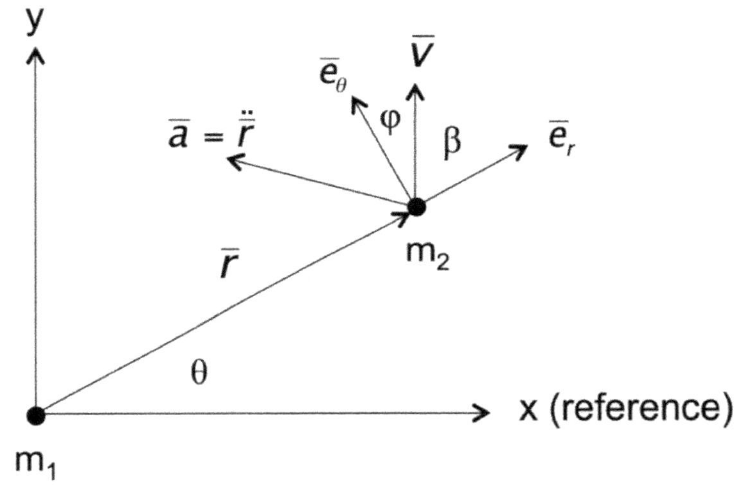


where

$$\bar{r} = r\,\bar{e}_r$$

$$\bar{v} = \dot{\bar{r}} = v_r\,\bar{e}_r + v_\theta\,\bar{e}_\theta = v\sin\varphi\,\bar{e}_r + v\cos\varphi\,\bar{e}_\theta$$

$$\bar{e}_k = \bar{e}_r \times \bar{e}_\theta$$

$$\bar{h} = \bar{r} \times \bar{v} = (r\,\bar{e}_r) \times (v\sin\varphi\,\bar{e}_r + v\cos\varphi\,\bar{e}_\theta) = rv\sin\varphi\ (\bar{e}_r \times \bar{e}_r) + r\cos(\bar{e}_r \times \bar{e}_\theta)$$

$$\bar{h} = rv\cos\varphi\,\bar{e}_k = rv\sin\beta\,\bar{e}_k$$

$\varphi$ = flight path angle

The magnitude of $\bar{h}$ is given as

$$h = |\bar{h}| = rv\cos\varphi = rv\sin\beta$$

As r and v change, the flight path angle, φ, must also change to keep h constant.

Since $\bar{h}$ is constant

$$\dot{\bar{h}} = 0$$

In polar coordinates

$$\bar{h} = \bar{r} \times \bar{v} = (r\,\bar{e}_r) \times (r\dot{\theta})\,\bar{e}_\theta = r^2\dot{\theta}\,\bar{e}_k$$

The magnitude of $\bar{h}$ is

$$h=|\bar{h}| = r^2\dot{\theta}$$

Energy Integral

Beginning with the Two-Body Relative Equations of Motion, take the scalar, i.e., dot, product of the velocity vector with Equation (3), as

$$\bar{r} \cdot \ddot{\bar{r}} = \bar{r} \cdot \left(\frac{-\mu}{r^3}\bar{r}\right)$$

which can be written

$$\frac{d}{dt}\left(\frac{\dot{\bar{r}} \cdot \dot{\bar{r}}}{2}\right) = \frac{d}{dt}\left(\frac{\mu}{r}\right)$$

Integrating gives

$$\frac{v^2}{2} = \frac{\mu}{r} + E$$

Solving for E shows

$$E = \frac{v^2}{2} - \frac{\mu}{r} \tag{6}$$

where

E =constant of integration equal to the total mechanical energy (scalar)

$T = \frac{v^2}{2}$ = kinetic energy

$V = -\frac{\mu}{r}$ = potential energy

such that

$$E = T + V$$

Equation (6) is known as the Energy Integral, which shows that the total mechanical energy of $m_2$ relative to $m_1$ is constant. Since E is a scalar quantity, it provides one integral for the Relative Two-Body Problem and two more integrals are still needed.

In summary, for the motion of $m_2$ about $m_1$, the Energy Integral requires:

- The sum of the kinetic energy and potential energy is constant.
- As r increases, v must decrease.
- As r decreases, v must increase.

Note that the potential energy will be negative $(-)$ since r is the distance from the center of $m_1$ to the center of $m_2$.

<u>Example 5.1</u>

The initial position and velocity of a satellite relative to Earth is given as

$$\bar{r}_1 = 5{,}200\,\bar{\imath} + 7{,}200\,\bar{\jmath}\ \text{km} \qquad \bar{v}_1 = -8.5\,\bar{\imath} + 7.9\,\bar{\jmath}\ \frac{\text{km}}{\text{s}}$$

The satellite changes its location under the gravitational attraction of Earth such that the new velocity is

$$\bar{v}_2 = 6.8\,\bar{\imath} + 1.3\,\bar{\jmath}\ \frac{\text{km}}{\text{s}}$$

Determine the position of the satellite and the corresponding flight path angle at this new location.

Solution:

$$\bar{h} = \bar{r}_1 \times \bar{v}_1 = (5{,}200\,\bar{\imath} + 7{,}200\,\bar{\jmath}) \times (-8.5\,\bar{\imath} + 7.9\,\bar{\jmath}) = 102{,}280{,}0\,\bar{k}\,\frac{km^2}{s}$$

Let $\bar{r}_2 = x\,\bar{\imath} + y\,\bar{\jmath}$

$$\bar{h} = \bar{r}_2 \times \bar{v}_2 = (x\,\bar{\imath} + y\,\bar{\jmath}) \times (\,6.8\,\bar{\imath} + 1.3\,\bar{\jmath}) = (1.3x - 6.8y)\,\bar{k}$$

$$r_1 = \sqrt{x_1^2 + y_1^2} = \sqrt{(5{,}200)^2 + (7{,}200)^2} = 8{,}881.4\ km$$

$$v_1 = \sqrt{v_{1_x}^2 + v_{1_y}^2} = \sqrt{(-8.5)^2 + (7.9)^2} = 11.60\ \frac{km}{s}$$

$$E = \frac{v_1^2}{2} - \frac{\mu}{r_1}$$

$$E = \frac{(11.60)^2}{2} - \frac{3.986 \times 10^5}{8{,}881.4} = 22.45\ \frac{km^2}{s^2}$$

$$v_2 = \sqrt{v_{2_x}^2 + v_{2_y}^2} = \sqrt{(6.8)^2 + (1.3)^2} = 6.92\ \frac{km}{s}$$

Since the energy of the satellite is constant

$$E = \frac{v_2^2}{2} - \frac{\mu}{r_2}$$

$$22.45 = \frac{(6.92)^2}{2} - \frac{3.986 \times 10^5}{r_2}$$

$$\Rightarrow r_2 = 263{,}282.3\ km$$

$$x^2 + y^2 = r_2^2 = (263{,}282.3)^2$$

Since angular momentum is also constant, equating the angular momentum terms shows

$$1.3x - 6.8y = 102{,}280.0$$

$$\Rightarrow x = 78{,}676.9 + 5.23y$$

Substituting gives

$$(78{,}676.9 + 5.23y)^2 + y^2 = (263{,}282.3)^2$$

$$6{,}190{,}058{,}225 + 823{,}117.9y + 28.36y^2 = 6.932 \times 10^{10}$$

$$28.36y^2 + 823{,}117.94y - 6.31275 \times 10^{10} = 0$$

Solving the quadratic equation above gives two solutions as

$$y = 34{,}847.7, -63{,}868.5 \text{ km} \implies x = 260{,}965.2, -255419.4 \text{ km}$$

$$\implies \bar{r}_2 = 260{,}965.2\,\bar{\imath} + 34{,}847.7\,\bar{\jmath}\,\text{km}, \ \underline{\text{or}}\ \bar{r}_2 = -255{,}419.4\,\bar{\imath} - 63{,}868.6\,\bar{\jmath}\,\text{km}$$

$$r_2 = \sqrt{(260{,}965.2)^2 + (34{,}847.7)^2} = 263{,}281.6 \text{ km}$$

$$\varphi = \cos^{-1}\left(\frac{h}{r_2 v_2}\right) = \cos^{-1}\left[\frac{102{,}280.0}{(263{,}281.6)(6.92)}\right]$$

$$\implies \varphi = 86.78^o$$

Example 5.2

An Earth satellite is tracked from a ground station and observed to have an altitude of 505 km and a velocity (expressed in polar coordinates) relative to Earth as

$$\bar{v} = 4.25\,\bar{e}_r + 7.75\,\bar{e}_\theta\ \frac{\text{km}}{\text{s}}$$

Calculate the following parameters for the satellite (relative to Earth): (a) speed; (b) flight-path angle; (c) energy; and (d) angular momentum.

Solution:

$$\text{(a)}\ v = \sqrt{v_r^2 + v_\theta^2} = \sqrt{(4.25)^2 + (7.75)^2}$$

$$\implies v = 8.84\ \frac{\text{km}}{\text{s}}$$

$$\text{(b)}\ \varphi = \tan^{-1}\left(\frac{v_r}{v_\theta}\right) = \tan^{-1}\left(\frac{4.25}{7.75}\right)$$

$$\implies \varphi = 28.74^o$$

(c) $r = R_E + (alt) = 6{,}378.1 + 505 = 6{,}883.1$ km

$$E = \frac{v^2}{2} - \frac{\mu}{r} = \frac{(8.84)^2}{2} - \frac{3.986\times10^5}{6{,}883.1}$$

$$\Rightarrow E = -18.84\ \frac{km^2}{s^2}$$

(d) $h = r^2\dot{\theta} = r(r\dot{\theta}) = r(v_\theta) = 6{,}883.1(7.75)$

$$\Rightarrow h = 53{,}344.0\ \frac{km^2}{s}$$

Example 5.3

An Earth satellite has a speed of 7.80 $\frac{km}{s}$ when its flight path angle is zero and the altitude is 821.9 km. Find the satellite's speed and flight path angle at an altitude of 1,541.9 km.

Solution:

$$r_1 = R_E + (alt)_1 = 6{,}378.1 + 821.9 = 7{,}200.0 \text{ km}$$

$$h = r_1 v_1 \cos\varphi_1 = (7{,}200.0)(7.80)\cos 0^o = 56{,}160.0\ \frac{km^2}{s}$$

$$E = \frac{v_1^2}{2} - \frac{\mu}{r_1} = \frac{(7.80)^2}{2} - \frac{3.986\times10^5}{7{,}200.0} = -24.9\frac{km^2}{s^2}$$

$$r_2 = R_E + (alt)_2 = 6{,}378.1 + 1{,}541.9 = 7{,}920.0 \text{ km}$$

$$E = \frac{v_2^2}{2} - \frac{\mu}{r_2}$$

$$v_2 = \sqrt{2\left(\frac{\mu}{r_2} + E\right)} = \sqrt{2\left(\frac{3.986\times10^5}{7{,}920.0} - 24.9\right)}$$

$$\Rightarrow v_2 = 7.13\ \frac{km}{s}$$

$$h = r_2 v_2 \cos\varphi_2$$

$$\varphi_2 = \cos^{-1}\left(\frac{h}{r_2 v_2}\right) = \cos^{-1}\left[\frac{56{,}160.0}{(7{,}920.0)(7.13)}\right]$$

$$\Rightarrow \varphi_2 = 6.00^o$$

## Problems

5.1 The initial position and velocity of a satellite relative to Earth is given as

$$\bar{r}_1 = 6{,}600\,\bar{\imath} + 4{,}000\,\bar{\jmath}\ \text{km} \qquad \bar{v}_1 = -4.5\,\bar{\imath} + 0.5\,\bar{\jmath}\ \frac{\text{km}}{\text{s}}$$

The satellite changes its location under the gravitational attraction of Earth such that the new velocity is

$$\bar{v}_2 = 6.0\,\bar{\imath} + 1.5\,\bar{\jmath}\ \frac{\text{km}}{\text{s}}$$

Determine the new position of the satellite and corresponding flight path angle at this new location.
(Ans. $\bar{r}_1 = 6{,}285.9\,\bar{\imath} - 1{,}978.5\,\bar{\jmath}$ km, or $\bar{r}_1 = -4{,}615.4\,\bar{\imath} - 4{,}703.8\,\bar{\jmath}$ km, $\varphi = 58.47^{\circ}$)

5.2 A rocket is fired in the vertical direction from Earth's surface and achieves a speed of $4.6\ \frac{\text{km}}{\text{s}}$ at an altitude of 160 km when the engine burns out. Neglecting the effects of Earth rotation and Earth's atmosphere
(a) determine the maximum altitude attained by this rocket
(b) find the speed of the burned-out rocket as it falls back to Earth and impacts the ground.
(Ans. (a) alt =1,533.0 km; (b) v = $4.9\ \frac{\text{km}}{\text{s}}$)

5.3 An Earth satellite is tracked from a ground station and observed to have an altitude of 652 km and a velocity (expressed in polar coordinates) relative to Earth as

$$\bar{v} = 3.72\,\bar{e}_r + 7.00\,e_\theta\ \frac{\text{km}}{\text{s}}$$

Calculate the following parameters for the satellite (relative to Earth):
(a) speed; (b) flight-path angle; (c) energy; (d) angular momentum
(Ans. (a) $v = 7.93\,\frac{\text{km}}{\text{s}}$; (b) $\varphi = 27.99^{\circ}$; (c) $E = -25.19\ \frac{\text{km}^2}{\text{s}^2}$; (d) $h = 49{,}280.7\,\frac{\text{km}^2}{\text{s}}$)

5.4 A projectile is launched from the surface of Earth with a speed of $3.6\,\frac{\text{km}}{\text{s}}$ and at an angle of 36° with the horizontal. Determine the maximum altitude achieved by the projectile.
(Ans. (alt) = 272.8 km)

5.5 For the launch scenario described in Problem 5.4, find the necessary launch velocity if the maximum altitude achieved is equal to $0.62R_E$ and the angle (measured from the center of Earth) between the launch point and impact point is 90°.
(Ans. $v = 7.76\ \frac{km}{s}$)

5.6 An Earth satellite has an altitude of 457 km, a flight path angle of 21.8°, and its angular momentum is given as $\bar{h} = 51{,}264.25\ \bar{e}_z\ \frac{km^2}{s}$. Determine the velocity vector in polar coordinates and calculate energy of the satellite.
(Ans. $\bar{v} = 3.00\ \bar{e}_r + 7.50\ \bar{e}_\theta\ \frac{km}{s}$, $E = -25.69\ \frac{km^2}{s^2}$)

5.7 The initial position and velocity of the satellite in Example 5.1 is given as

$$\bar{r}_1 = 5{,}200\,\bar{\imath} + 7{,}200\,\bar{\jmath}\ km \qquad \bar{v}_1 = -8.5\,\bar{\imath} + 7.9\,\bar{\jmath}\ \frac{km}{s}$$

The satellite changes its location under the gravitational attraction of Earth such that the position vector becomes

$$\bar{r}_2 = -5{,}400\,\bar{\imath} + 7{,}400\,\bar{\jmath}\ km$$

Determine the velocity vector of the satellite at its new location.
(Ans. $\bar{v}_2 = -7.43\,\bar{\imath} - 8.76\,\bar{\jmath}\ \frac{km}{s}$, or $\bar{v}_2 = -10.61\,\bar{\imath} - 4.40\,\bar{\jmath}\ \frac{km}{s}$)

5.8 Consider a projectile launched from the surface of Earth having velocity, v, an angle of $\alpha$ with the horizontal, and that it reaches a maximum altitude of $H_{max}$.
(a) For the case of $v = 3.45\ \frac{km}{s}$ and $H_{max} = 350$ km, determine $\alpha$.
(b) For the case of $v = 3.65\ \frac{km}{s}$ and $\alpha = 38°$, calculate $H_{max}$.
(c) If $\alpha = 42°$ and $H_{max} = 375$ km, compute the launch velocity required.
(Ans. (a) $\alpha = 44.72°$; (b) $H_{max} = 309.8$ km; (c) $v = 3.70\ \frac{km}{s}$)

# Module 6: Kepler's Laws

Johannes Kepler published his first two laws dealing with planetary motion in 1609, in the publication entitled, *Astronomia Nova*, or *The New Astronomy*. He published a third law dealing with planetary motion in a publication entitled, *Harmonicies Mundi*, or *The Harmony of the World*, in 1619.

Kepler's Laws are based on the measurements of planetary positions made by the Danish astronomer Tycho Brahe. These laws provide a description of how planets move within the solar system and are empirical in nature. They describe how planets move but don't explain why they move in the way they do.

Kepler is credited with reformulating the science of astronomy through his laws and each one plays an important role in our understanding of two-body motion. Each of Kepler's Laws provide important information in describing the orbit of planets about the Sun.

## Kepler's First Law

This law states: *The orbit of each planet is an ellipse, with the Sun at a focus.*

In order to understand the implications of this law, consider the elliptical orbit shown below.

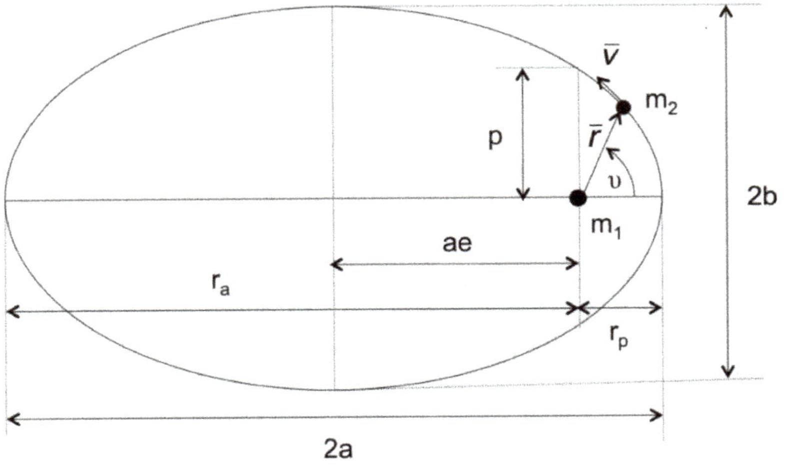

where

$r = |\bar{r}| =$ distance between $m_1$ and $m_2$
$a =$ semi-major axis, or a measure of the size of the ellipse
$b =$ semi-minor axis
$p =$ parameter
$e =$ eccentricity, or a measure of the shape of the ellipse
$\nu =$ true anomaly, the angular position of $m_2$ from periapsis in the direction of travel

The geometry of an ellipse is described by the expression

$$r = \frac{p}{1 + e\cos\nu} = \frac{a(1-e^2)}{1 + e\cos\nu} \tag{7}$$

Equation (7) is based on the geometry of the ellipse only, not dynamics.

Other useful expressions regarding the ellipse are

$$r_p = a(1 - e)$$

$$r_a = a(1 + e)$$

$$a = \frac{r_p + r_a}{2}$$

$$e = \frac{r_a - r_p}{r_a + r_p}$$

$$p = a(1 - e^2)$$

$$b = a\sqrt{1 - e^2}$$

For elliptical motion about any body, the following general terms apply

$r_p =$ closest point to $m_1 =$ periapsis/pericenter
$r_a =$ farthest point from $m_1 =$ apoapsis/apocenter

For specific primary bodies, or $m_1$, the following terms apply

| | | |
|---|---|---|
| Sun | $\Rightarrow$ | perihelion/aphelion |
| Earth | $\Rightarrow$ | perigee/apogee |
| Moon | $\Rightarrow$ | perilune/apolune |
| Jupiter | $\Rightarrow$ | perijove/apojove |
| Star | $\Rightarrow$ | periastor/apoastor |

Kepler's Second Law

This law states: *The radius vector joining the Sun to a planet sweeps out equal areas in equal intervals of time.*

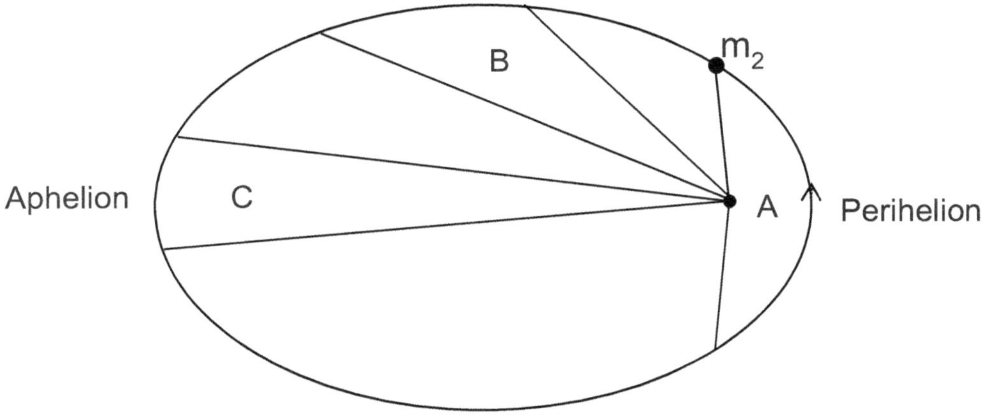


This law indicates that the orbit of planet $m_2$ about the Sun takes the same amount of time to sweep out the equal areas of A, B, and C as shown above, or

$$\text{If}\ \ \Delta t_A = \Delta t_B = \Delta t_C \ \Rightarrow\ A_A = A_B = A_C$$

This fact can be used to obtain a very important relationship regarding two-body relative motion. Consider the orbital path of $m_2$ through the angle $d\theta$, which sweeps out area dA in time dt, as shown below.

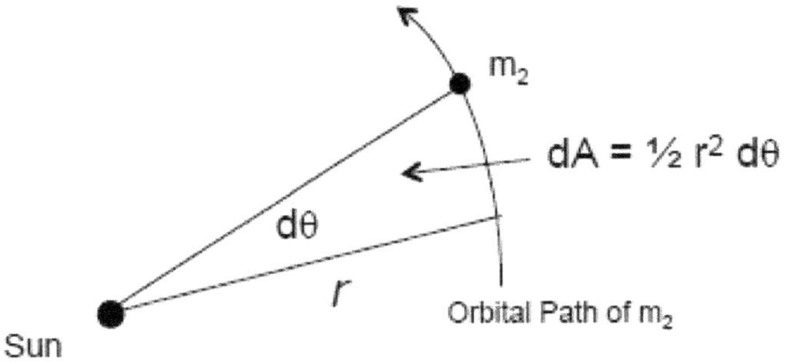


Time can be introduced by writing the expression for the areal velocity

$$\frac{dA}{dt} = \frac{r^2}{2}\left(\frac{d\theta}{dt}\right) = \frac{1}{2}\left(r^2\dot{\theta}\right) = \frac{1}{2}h = \text{Constant}$$

This expression can be written

$$dA = \frac{h}{2}dt$$

Integrating dA over the area of an ellipse gives

$$2\int_{Ellipse}(dA) = \int_{t_0}^{t} hdt$$

$$2A_{Ellipse} = hT$$

$$2\pi ab = hT$$

$$2\pi a\sqrt{a^2(1-e^2)} = \sqrt{\mu a(1-e^2)}\,T$$

$$2\pi\sqrt{a^3} = \sqrt{\mu}\,T$$

$$\Rightarrow \quad T = 2\pi\sqrt{\frac{a^3}{\mu}} \qquad (8)$$

Equation (8) is an expression for the 'orbital period', T, of a circular or elliptical orbit of $m_2$ about $m_1$. The period is only a function of the orbit size, a, and gravitational parameter, $\mu$.

Kepler's Third Law

This law can be written as, *The square of the period of a planet is proportional to the cube of its mean distance from the Sun.*

which states

$$T^2 \sim a^3$$

Kepler's Second Law provides

$$T = 2\pi\sqrt{\frac{a^3}{\mu}}$$

$$\Rightarrow T^2 = (4\pi^2)\frac{a^3}{\mu}$$

which proves Kepler's Third Law.

The ten integrals already found for the solution to the inertial two-body problem provide the following information:

- The Center-of-Mass Integrals established a relative coordinate system for the motion of $m_2$ relative to $m_1$ (six integrals). These integrals were used to transform the Inertial Two-Body Problem to the Relative Two-Body Problem.

- The Angular Momentum Integrals showed that the angular momentum of $m_2$ relative to $m_1$ must be constant, both in magnitude and direction (three integrals).

- The Energy Integral showed the total mechanical energy (kinetic energy plus potential energy) of $m_2$ relative to $m_1$ must be constant (one integral).

Two more integrals remain to be found in order to obtain a complete solution to the Relative Two-Body Problem. These integrals are presented in Modules 7 and 9.

## The Vis-Viva Equation

Another important relationship which is very useful in many types of orbital mechanics calculations can now be derived. It provides a method to determine the velocity of $m_2$ at all points in its orbit. This relationship can be obtained beginning with the Energy Integral

$$E = \frac{v^2}{2} - \frac{\mu}{r} = \text{Constant} \tag{9}$$

The energy of $m_2$ at periapsis can be evaluated using the relationships

$$r_p = a(1 - e)$$

$$v_p = \frac{h}{r_p}$$

$$h^2 = \mu a(1 - e^2)$$

Substituting these expressions into Equation (9) yields an alternate expression for energy as

$$E = \frac{-\mu}{2a} \tag{10}$$

This result shows that the energy of $m_2$ is only a function of the semi-major axis. Equating Equations (9) and (10) and solving for v gives

$$v = \sqrt{\mu\left(\frac{2}{r} - \frac{1}{a}\right)} \quad (11)$$

Equation (11) is known as the vis-viva equation and is used extensively in orbital transfer and mission planning calculations.

<u>Example 6.1</u>

A satellite has perigee and apogee altitudes of 250 km and 42,000 km, respectively. Calculate the orbital period, eccentricity, and the maximum speed of the satellite.

Solution:

$$r_p = R_E + 250 = 6{,}378.1 + 250 = 6{,}628.1 \text{ km}$$

$$r_a = R_E + 42{,}000 = 6{,}378.1 + 42{,}000 = 48{,}378.1 \text{ km}$$

$$a = \frac{r_p + r_a}{2} = \frac{6{,}628.1 + 48{,}378.1}{2} = 27{,}503.1 \text{ km}$$

$$T = 2\pi\sqrt{\frac{a^3}{\mu}} = 2\pi\sqrt{\frac{(27{,}503.1)^3}{3.986 \times 10^5}}$$

$$\Rightarrow \quad T = 45{,}392.46 \text{ sec} = 756.54 \text{ min} = 12.61 \text{ hrs}$$

$$r_a = a(1 + e)$$

$$e = \frac{r_a}{a} - 1 = \frac{48{,}378.1}{27{,}503.1} - 1$$

$$\Rightarrow \quad e = 0.759$$

$$v_p = \sqrt{\mu\left(\frac{2}{r_p} - \frac{1}{a}\right)} = \sqrt{3.986 \times 10^5 \left(\frac{2}{6{,}628.1} - \frac{1}{27{,}503.1}\right)}$$

$$\Rightarrow \quad v_p = 10.29 \ \frac{\text{km}}{\text{s}}$$

Example 6.2

A satellite in a polar Earth orbit comes within 150 km of the North Pole at its closest approach. If the satellite passes over the pole once every 90 minutes, determine the eccentricity of its orbit.

Solution:

$$T = 90 \text{ min} = 5{,}400 \text{ sec}$$

$$r_p = R_E + 150.0 = 6{,}378.1 + 150 = 6{,}528.1 \text{ km}$$

$$T = 2\pi\sqrt{\frac{a^3}{\mu}}$$

$$a = \sqrt[3]{\frac{\mu T^2}{4\pi^2}} = \sqrt[3]{\frac{(3.986\times10^5)(5{,}400)^2}{4\pi^2}} = 6{,}652.6 \text{ km}$$

$$r_p = a(1 - e)$$

$$e = 1 - \frac{r_p}{a} = 1 - \frac{6{,}528.1}{6{,}652.6}$$

$$\Rightarrow \quad e = 0.019$$

Example 6.3

A spacecraft has an angular momentum of 70,000 $\frac{km^2}{s}$ and energy equal to $-10.0\frac{km^2}{s^2}$. Find the altitude at the perigee and apogee points of its orbit.

Solution:

$$E = \frac{-\mu}{2a}$$

$$a = \frac{-\mu}{2E} = \frac{-3.986\times10^5}{2(-10.0)} = 19{,}930.0 \text{ km}$$

$$\frac{h^2}{\mu} = a(1 - e^2)$$

$$e = \sqrt{1 - \frac{h^2}{\mu a}} = \sqrt{1 - \frac{(70{,}000)^2}{3.986\times10^5(19{,}930.0)}} = 0.619$$

$r_p = a(1 - e) = 19{,}930.0(1 - 0.619) = 7{,}592.9$ km

$(alt)_p = r_p - R_E = 7{,}592.9 - 6{,}378.1$

$\Rightarrow (alt)_p = 1{,}214.8$ km

$r_a = a(1 + e) = 19{,}930.0(1 + 0.619) = 32{,}266.7$ km

$(alt)_a = r_a - R_E = 32{,}266.7 - 6{,}378.1$

$\Rightarrow (alt)_a = 25{,}888.6$ km

Problems

6.1 A spacecraft is launched with a speed of 8.0 $\frac{km}{s}$ parallel to Earth's surface at an altitude of 600 km. Determine the period of the spacecraft in minutes.
(Ans. $T = 117.21$ min)

6.2 A geostationary satellite is one which orbits Earth directly above the equator at a speed such that it always remains above the same point on Earth. Calculate
(a) the altitude and speed of a geostationary Earth satellite
(b) the maximum latitude and the percentage of Earth's surface visible from a geostationary Earth satellite.
(Ans. (a) (alt) $= 35{,}786$ km, $v = 3.075 \frac{km}{s}$; (b) $\phi_N = 81.3^o$, 42.4% of Earth's surface is visible)

6.3 A spacecraft at a perigee altitude of 500 km has a speed of 10.0 $\frac{km}{s}$. Compute the spacecraft's speed, flight path angle, and altitude when it's located at a true anomaly of 120°.
(Ans. $v = 5.19 \frac{km}{s}$, $\varphi = 44.6^o$, alt $= 12{,}247.5$ km)

6.4 A satellite has perigee and apogee altitudes of 350 km and 21,000 km, respectively. Calculate the semi-major axis, orbital period, eccentricity, and maximum speed of the satellite.
(Ans. $a = 17{,}053.1$ km, $T = 369.37$ min, $e = 0.604$, $v_p = 9.75 \frac{km}{s}$)

6.5 A satellite has an orbital period of 210 minutes and a speed at apogee of 5.02 $\frac{km}{s}$. Determine the altitude and speed of the satellite at perigee.
(Ans. $(alt)_p = 3{,}569.7$ km, $v_p = 6.79 \frac{km}{s}$)

6.6 A spacecraft is in an Earth orbit having a = 9,800 km and e = 0.18. Calculate the following quantities:

(a) the maximum speed during its orbit

(b) the angular momentum

(c) the speed at a true anomaly of 210°

(d) the true anomaly of the minor axis of the orbit.

(Ans. (a) $v_p = 7.65\frac{km}{s}$, $h = 61{,}475.4\ \frac{km^2}{s}$, $v = 5.50\ \frac{km}{s}$, $\nu = 100.37^o$)

6.7 Consider the mass of the moon to be $\frac{1}{81.28}$ times the mass of Earth, and that the period of the moon's orbit about Earth is 27.3217 days. Determine the semi-major axis of the moon's orbit.

(Ans. a = 384,748.2 km)

6.8 A U.S. Air Force Space Command early warning defense system detects a foreign-launched ballistic missile. At one point, the missile's orbital parameters are

speed = $9.0\ \frac{km}{s}$ altitude = 1,122 km

flight-path angle = 25° eccentricity = 0.636

Find the semi-major axis and the true anomaly at the location where the missile will impact the surface of Earth.

(Ans. a = 15,759.9 km, $\nu = 317.93^o$)

(This page was intentionally left blank.)

# Module 7: The Trajectory Equation

Kepler's First Law determined that the orbit of $m_2$ about $m_1$ follows an elliptical path. This conclusion was based on empirical observational data and geometry only. Isaac Newton introduced dynamics into the solution of the Two-Body Problem in 1687 in *Principia.*

Consider the Relative Two-Body Problem in polar coordinates in the orbital plane as shown in this figure

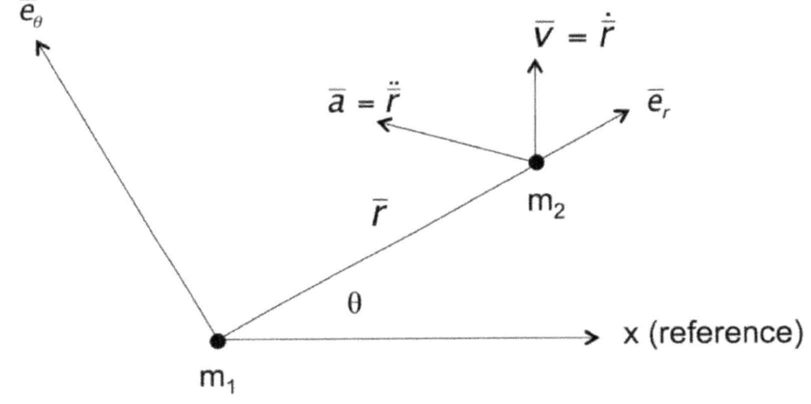


where the unit vectors, $\bar{e}_r$ and , $\bar{e}_\theta$ change directions as $m_2$ moves within the orbital plane.

Writing the position vector and taking the time derivative shows

$$\bar{r} = r\,\bar{e}_r$$

$$\dot{\bar{r}} = \frac{d\bar{r}}{dt} = \bar{v} = \frac{d}{dt}(r\,\bar{e}_r) = \dot{r}\,\bar{e}_r + r\,\dot{\bar{e}}_r$$

$$\dot{\bar{e}}_r = \frac{d\bar{e}_r}{dt} = \frac{d\bar{e}_r}{d\theta}\frac{d\theta}{dt} = \dot{\theta}\,\frac{d\bar{e}_r}{d\theta}$$

$$d\bar{e}_r = \bar{e}_\theta d\theta$$

$$\dot{\bar{e}}_r = \dot{\theta}\,\bar{e}_\theta$$

$$\Rightarrow \quad \bar{v} = \dot{r}\,\bar{e}_r + r\dot{\theta}\,\bar{e}_\theta$$

Also, the acceleration can also be obtained by differentiation as

$$\ddot{\bar{r}} = \bar{a} = \frac{d\bar{v}}{dt} = \frac{d}{dt}(\dot{r}\,\bar{e}_r + r\dot{\theta}\,\bar{e}_\theta) = \ddot{r}\,\bar{e}_r + \dot{r}\,\dot{\bar{e}}_r + \dot{r}\dot{\theta}\,\bar{e}_\theta + r\ddot{\theta}\,\bar{e}_\theta + r\dot{\theta}\,\dot{\bar{e}}_\theta$$

$$\ddot{\bar{r}} = \ddot{r}\,\bar{e}_r + \dot{r}\dot{\theta}\,\bar{e}_\theta + \dot{r}\dot{\theta}\,\bar{e}_\theta + r\ddot{\theta}\,\bar{e}_\theta + r\dot{\theta}\,\dot{\bar{e}}_\theta$$

$$\ddot{\bar{r}} = \ddot{r}\,\bar{e}_r + (2\dot{r}\dot{\theta} + r\ddot{\theta})\,\bar{e}_\theta + r\dot{\theta}\,\dot{\bar{e}}_\theta$$

Since

$$\dot{\bar{e}}_\theta = \frac{d\bar{e}_\theta}{dt} = \frac{d\bar{e}_\theta}{d\theta}\frac{d\theta}{dt} = \dot{\theta}\frac{d\bar{e}_\theta}{d\theta}$$

$$d\bar{e}_\theta = -\bar{e}_r\,d\theta$$

$$\dot{\bar{e}}_\theta = -\dot{\theta}\,\bar{e}_r$$

$$\Rightarrow\ \bar{a} = (\ddot{r} - r\dot{\theta}^2)\,\bar{e}_r + (2\dot{r}\dot{\theta} + r\ddot{\theta})\,\bar{e}_\theta$$

The Two-Body Relative Equations of Motion for $m_2$ can then be written

$$\bar{a} = \ddot{\bar{r}} = \frac{-\mu\bar{r}}{r^3} = \frac{-\mu}{r^3}(r\,\bar{e}_r)$$

$$\Rightarrow\ (\ddot{r} - r\dot{\theta}^2)\,\bar{e}_r + (2\dot{r}\dot{\theta} + r\ddot{\theta})\,\bar{e}_\theta = \frac{-\mu}{r^2}\,\bar{e}_r$$

The acceleration component in the radial direction is

$$(\ddot{r} - r\dot{\theta}^2)\,\bar{e}_r = \frac{-\mu}{r^2}\,\bar{e}_r$$

$$\Rightarrow\ \ddot{r} - \frac{r^4}{r^3}\dot{\theta}^2 = \frac{-\mu}{r^2}$$

Since

$$h = r^2\dot{\theta}$$

$$\Rightarrow\ \ddot{r} - \frac{h^2}{r^3} = \frac{-\mu}{r^2}$$

The above equation can be integrated using variable transformation.

Define

$$r = \frac{1}{s}$$

$$\dot{r} = \frac{dr}{dt} = \frac{dr}{ds}\frac{ds}{d\theta}\frac{d\theta}{dt}$$

Such that

$$\frac{dr}{ds} = \frac{d\left(\frac{1}{s}\right)}{ds} = \frac{-1}{s^2} = -r^2$$

$$\frac{d\theta}{dt} = \dot{\theta} = \frac{h}{r^2}$$

$$\Longrightarrow \quad \dot{r} = -r^2\left(\frac{ds}{d\theta}\right)\left(\frac{h}{r^2}\right) = -h\left(\frac{ds}{d\theta}\right)$$

This equation can be written as

$$\ddot{r} = \frac{d\dot{r}}{dt} = \left(\frac{d\dot{r}}{d\theta}\right)\left(\frac{d\theta}{dt}\right) = \frac{d}{d\theta}\left(-h\frac{ds}{d\theta}\right)\left(\frac{d\theta}{dt}\right) = -h\left(\frac{d^2s}{d\theta^2}\right)\dot{\theta} = -h\left(\frac{r^2\dot{\theta}}{r^2}\right)\left(\frac{d^2s}{d\theta^2}\right) = -\left(\frac{h^2}{r^2}\right)\left(\frac{d^2s}{d\theta^2}\right)$$

$$\Longrightarrow \quad \ddot{r} = -h^2s^2\left(\frac{d^2s}{d\theta^2}\right)$$

Substituting into the Two-Body Relative Equations of Motion gives

$$-h^2s^2\left(\frac{d^2s}{d\theta^2}\right) - \frac{h^2}{r^3} = \frac{-\mu}{r^2}$$

$$-h^2s^2\left(\frac{d^2s}{d\theta^2}\right) - h^2s^3 = -\mu s^3$$

$$\Longrightarrow \quad \left(\frac{d^2s}{d\theta^2}\right) + s = \frac{-\mu}{h^2}$$

This is the equation for a 'harmonic oscillator'. The solution to the inhomogeneous, linear differential equation is well-known as

$$s = \frac{\mu}{h^2} + \cos(\theta - \omega) = \frac{1}{r}$$

$$\Longrightarrow \quad r = \frac{\left(\frac{h^2}{\mu}\right)}{1+A\left(\frac{h^2}{\mu}\right)\cos(\theta-\omega)} \tag{12}$$

where

$h$ = angular momentum
$A$ = constant of integration
$\nu = \theta - \omega$

Comparing Equations (7) and (12) shows the connection between geometry and dynamics as

(geometry ≡ dynamics)

$$r = \frac{p}{1+e\cos\nu} = \frac{a(1-e^2)}{1+e\cos\nu} = \frac{\left(\frac{h^2}{\mu}\right)}{1+A\left(\frac{h^2}{\mu}\right)\cos(\theta-\omega)}$$

Equating terms gives

$$\frac{h^2}{\mu} = a(1-e^2) = p$$

$$e = \frac{Ah^2}{\mu}$$

$$\nu = \theta - \omega$$

The following equations provide one integral and are known together as the Trajectory Equation, which provides the important result that the motion of $m_2$ relative to $m_1$ will be a 'conic section' defined by a and e.

$$r = \frac{a(1-e^2)}{1+e\cos\nu}$$

$$a(1-e^2) = \frac{h^2}{\mu} = p \tag{13}$$

Equations (13) also provide a way of calculating the distance between $m_1$ and $m_2$ at all points in the orbit, which verifies Kepler's First Law. More importantly, it proves that the motion of $m_2$ is based on dynamics.

One more integral is needed to obtain a complete solution to the Relative Two-Body Problem. This final integral is presented in Module 9.

Example 7.1

A satellite has a perigee altitude of 500 km and a speed of 10.0 $\frac{km}{s}$. Find the eccentricity, energy, flight path angle, and altitude when it's true anomaly is 120°.

Solution:

$$r_p = R_E + (alt)_p = 6{,}378.1 + 500 = 6{,}878.1 \text{ km}$$

$$h = r_p v_p = (6{,}878.1)(10.0) = 68{,}781.0 \ \frac{km^2}{s^2}$$

$$r_p = \frac{\frac{h^2}{\mu}}{1+e\cos 0^o} = \frac{\frac{h^2}{\mu}}{1+e}$$

$$e = \frac{h^2}{r_p\mu} - 1 = \frac{(68{,}781.0)^2}{(6{,}878.1)(3.986\times10^5)} - 1$$

$$\Rightarrow \quad e = 0.726$$

$$E = \frac{v_p^2}{2} - \frac{\mu}{r_p} = \frac{(10.0)^2}{2} - \frac{3.986\times10^5}{6{,}878.1}$$

$$\Rightarrow \ E = -7.95 \ \frac{km^2}{s^s}$$

at $\nu = 120^o$:

$$r = \frac{\frac{h^2}{\mu}}{1+e\cos\nu} = \frac{\frac{(68{,}781.0)^2}{3.986\times10^5}}{1+0.726\cos 120^o} = 18{,}632.0 \text{ km}$$

$$v = \sqrt{\left[2\left(E+\frac{\mu}{r}\right)\right]} = \sqrt{\left[2\left(-7.95+\frac{3.986\times10^5}{18{,}632.0}\right)\right]} = 5.19 \ \frac{km}{s}$$

$$h = rv\cos\varphi$$

$$\varphi = \cos^{-1}\left(\frac{h}{rv}\right) = \cos^{-1}\left(\frac{68{,}781.0}{18632.0(5.19)}\right)$$

$$\Rightarrow \quad \varphi = 44.7^o$$

$$(alt) = r - R_E = 18{,}632.0 - 6{,}378.1$$

$$\Rightarrow (alt) = 12{,}253.9 \text{ km}$$

Example 7.2

A satellite orbiting Earth at a radius of 8,000 km has a speed of 7.5 $\frac{km}{s}$ and a flight-path angle of $10^o$. Determine the eccentricity of the orbit and the true anomaly of the satellite at the specified point in its orbit.

Solution:

$$E = \frac{v^2}{2} - \frac{\mu}{r} = \frac{(7.5)^2}{2} - \frac{3.986\times10^5}{8{,}000} = -21.7\ \frac{km^2}{s^2}$$

$$E = \frac{-\mu}{2a}$$

$$a = \frac{-\mu}{2E} = \frac{-3.986\times10^5}{2(-21.7)}$$

$$\Rightarrow a = 9{,}184.3\ km$$

$$h = rv\cos\varphi = 8{,}000(7.5)\cos 10^o = 59{,}088.47\ \frac{km^2}{s}$$

$$\frac{h^2}{\mu} = a(1 - e^2)$$

$$e = \sqrt{1 - \frac{h^2}{a\mu}} = \sqrt{1 - \frac{(59{,}088.47)^2}{9{,}184.3(3.986\times10^5)}}$$

$$\Rightarrow e = 0.215$$

$$r = \frac{\frac{h^2}{\mu}}{1 + e\cos\nu}$$

$$\nu = \cos^{-1}\left\{\frac{1}{e}\left[\frac{h^2}{\mu r} - 1\right]\right\} = \cos^{-1}\left\{\frac{1}{0.215}\left[\frac{(59{,}088.47)^2}{3.986\times10^5(8{,}000)} - 1\right]\right\}$$

$$\Rightarrow \nu = 63.8^o$$

Example 7.3

At two different points of an Earth orbit, a satellite's altitude and true anomaly are given as 1,545 km at $\nu_1 = 126^o$ and 852 km at $\nu_2 = 58^o$. Determine the eccentricity, semi-major axis, altitude at perigee, and orbital period of this satellite.

Solution:

$$r_1 = R_E + (alt)_1 = 6{,}378.1 + 1{,}545 = 7{,}923.1 \text{ km}$$

$$r_2 = R_E + (alt)_2 = 6{,}378.1 + 852 = 7{,}230.1 \text{ km}$$

At point 1,

$$r_1 = \frac{a(1-e^2)}{1+e\cos\nu_1}$$

$$7{,}923.1 = \frac{a(1-e^2)}{1+e\cos 126^o}$$

$$\Rightarrow (1-0.588\ e)(7{,}923.1) = a(1-e^2)$$

At point 2,

$$r_2 = \frac{a(1-e^2)}{1+e\cos\upsilon_2}$$

$$7{,}230.1 = \frac{a(1-e^2)}{1+e\cos 58^o}$$

$$\Rightarrow (1+0.530\ e)(7{,}230.1) = a(1-e^2)$$

Equating the terms above shows

$$(1-0.588\ e)(7{,}923.1) = (1+0.530\ e)(7{,}230.1)$$

$$\Rightarrow e = 0.082$$

$$a = \frac{(1+0.530\ e)(7{,}230.1)}{1-e^2} = \frac{[1+0.530\ (0.082)](7{,}230.1)}{1-(0.082)^2}$$

$$\Rightarrow a = 7{,}595.4 \text{ km}$$

$(alt)_p = r_p - R_E = a(1-e) = 7{,}595.4(1-0.082) - 6{,}378.1$

$\Longrightarrow (alt)_p = 594.5 \text{ km}$

$$T = 2\pi\sqrt{\frac{a^3}{\mu}} = 2\pi\sqrt{\frac{(7{,}595.4)^3}{3.986\times10^5}}$$

$\Longrightarrow T = 6{,}587.8 \text{ sec} = 109.80 \text{ min} = 1.83 \text{ hrs}$

## Problems

7.1 A vehicle is launched from the surface of Earth with a velocity of 4.8 $\frac{km}{s}$ and a flight path angle of 42°. Calculate the maximum altitude above Earth's surface achieved by this vehicle.
(Ans. 726.8 km)

7.2 An Earth satellite is tracked from a ground station and observed to have an altitude of 662 km and a velocity (expressed in polar coordinates) as

$$\bar{v} = -4.00\,\bar{e}_r + 6.50\,\bar{e}_\theta\,\frac{km}{s}$$

Determine the values for the following orbital parameters of the satellite
(a) semi-major axis
(b) parameter
(c) eccentricity
(d) true anomaly.
(Ans. (a) a =7,249.9 km; (b) 5,253.5 km; (c) e = 0.525; (d) ν = 240.1°)

7.3 Find an expression for the true anomaly of the points on an elliptical orbit where the speed equals the speed on a circular orbit having the same value of r.
(Ans. $\nu = \cos^{-1}(-e)$)

7.4 A satellite is launched into Earth orbit with a speed of 10.2 $\frac{km}{s}$ at an altitude of 1,000 km and having a flight-path angle of 16°. Compute the true anomaly of the launch point and the orbital period of the resulting orbit.
(Ans. ν = 33.2°, T = 86.7 hr)

7.5 A spacecraft has a perigee radius of 10,000 km and an apogee radius of 100,000 km. Calculate the following quantities

(a) eccentricity
(b) semi-major axis
(c) orbital period
(d) energy
(e) true anomaly when the spacecraft altitude is 10,000 km
(f) velocity components $v_r$ and $v_\theta$ at the point found in (e)
(g) the speed at apogee and perigee.

(Ans. (a) $e = 0.8182$; (b) $a = 55{,}000$ km; (c) $T = 35.66$ hr; (d) $E = -3.624\,\frac{km^2}{s^2}$; (e) $\nu = 82.3^o$; (f) $v_r = 3.80\,\frac{km}{s}, v_\theta = 5.20\,\frac{km}{s}$; (g) $v_a = 0.85\,\frac{km}{s}, v_p = 8.51\,\frac{km}{s}$)

7.6 Data for an Earth satellite is given as

altitude = 1,000 km at $\nu = 40^o$ altitude – 2,000 km at $\nu = 150^o$

Determine the following quantities

(a) eccentricity
(b) altitude at perigee
(c) semi-major axis.

(Ans. (a) e – 0.078; (b) $(alt)_p = 875.9$ km; (c) $a = 7{,}863.2$ km)

7.7 A foreign launched vehicle was detected having a speed of $8.9\,\frac{km}{s}$ at an altitude of 1,100 km and a flight path angle of $20^o$. Find the eccentricity and radius at perigee.
(Ans. $e = 0.57$, $r_p = 6.247.8$ km)

7.8 The U.S. Air Force Space Command early warning defense system detects a foreign-launched ballistic missile. At one point, the missile's orbital parameters are found to be

speed = $9.0\,\frac{km}{s}$ altitude = 1,122 km flight-path angle = $25^o$

Calculate the true anomaly of the point where the missile will impact Earth's surface.
(Ans. $\nu = 317.8^o$)

(This page was intentionally left blank.)

# Module 8: Conic Section Motion

The Trajectory Equation provided by Equations (13) is applicable to all conic sections. It proves that the motion of $m_2$ about $m_1$ can only be circular, elliptical, parabolic, or hyperbolic and no other types of motion are possible. Properties of conic section motion are provided as follows:

1. Conic sections, i.e., circles, ellipses, parabolas, and hyperbolas, represent the only possible paths of $m_2$ about $m_1$.

2. The focus of the orbit must be at the center of the central body, $m_1$.

3. The energy of $m_2$ relative to $m_1$ is constant.

4. The angular momentum of $m_2$ relative to $m_1$ is constant.

5. The orbital plane defined by the motion of $m_2$ relative to $m_1$ is fixed in space.

A graphical representation of the four conic sections is provided in the figure below.

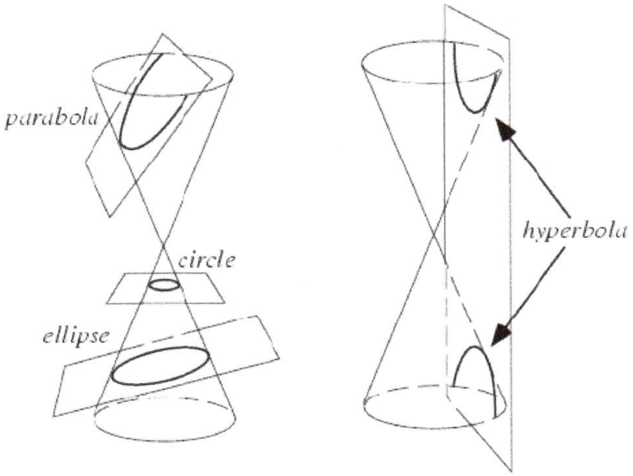


The following parameters are used to describe and classify the possible types of conic section motion:

$a$ = semi-major axis, a measure of the size of the conic section
$e$ = eccentricity, a measure of the shape of the conic section
$E = \frac{-\mu}{2a}$, the energy of $m_2$ in its orbit about $m_1$

Also, the quantity $h^2 = \mu a(1 - e^2) \geq 0$, is mathematically required.

The following table summarizes the values of these parameters for each type of conic section:

| Conic section | Semi-major axis | Energy | Eccentricity |
|---|---|---|---|
| circle | $0 < a < \infty$ | $E < 0$ | $e = 0$ |
| ellipse | $0 < a < \infty$ | $E < 0$ | $0 < e < 1$ |
| parabola (rare) | $a = \infty$ | $E = 0$ | $e = 1$ |
| hyperbola | $-\infty < a < 0$ | $E > 0$ | $e > 1$ |

As e gets larger, an ellipse becomes flatter and more elongated as shown in the figure

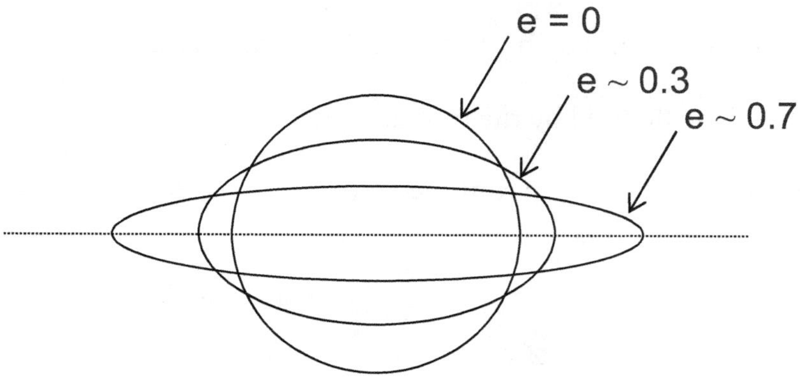


And as $e \Longrightarrow 1$, $a \Longrightarrow \infty$ and $E \Longrightarrow 0$.

The properties of each type of conic section motion are presented in the figures below.

Circular motion ($0 < a < \infty$, $E < 0$, $e = 0$, $r = a = \text{constant}$)

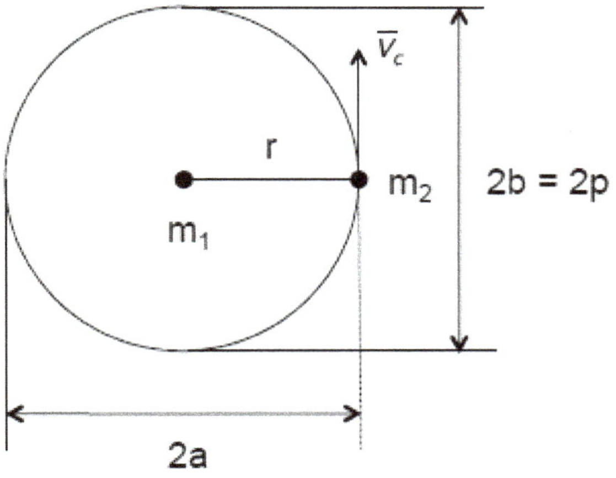


The circular velocity is given as

$$v_c = \sqrt{\frac{\mu}{a}}$$

As r gets larger, $v_c$ gets smaller.

- for a satellite in low-Earth orbit (LEO) $v_c \approx 7.5 \, \frac{km}{s}$
- for the Moon's orbit about Earth $v_c \approx 1.0 \, \frac{km}{s}$

Elliptical motion ($0 < a < \infty$, $E < 0$, $0 < e < 0$)

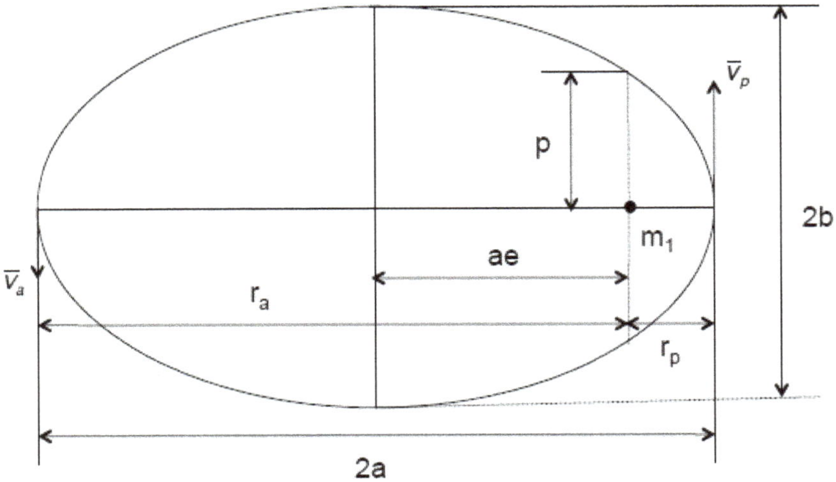

At periapsis: $r_p = a(1-e) \Rightarrow v_p = \sqrt{\frac{\mu(1+e)}{a(1-e)}}$

At apoapsis: $r_a = a(1+e) \Rightarrow v_a = \sqrt{\frac{\mu(1-e)}{a(1+e)}}$

At all other locations

$$r = \frac{a(1-e^2)}{1+e\cos\nu} \Rightarrow v_a < v < v_p$$

where

$$v = \sqrt{\mu\left(\frac{2}{r} - \frac{1}{a}\right)}$$

<u>Parabolic motion</u> ($a = \infty$, $E = 0$, $e = 1$, which rarely exists)

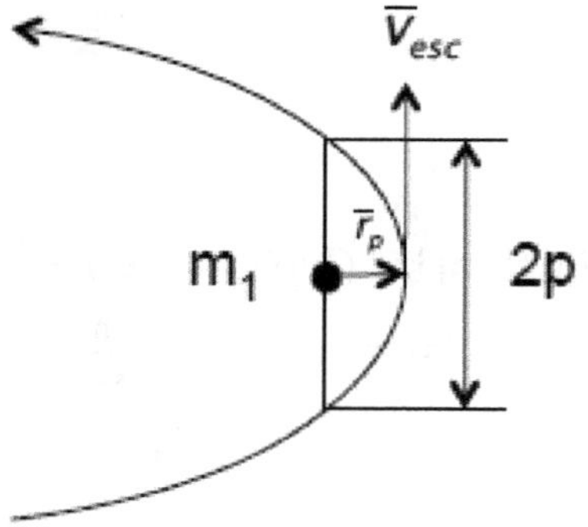


where

$$p = 2r_p$$

$$v_{esc} = \sqrt{\frac{2\mu}{r_p}} = \sqrt{2}\,v_c$$

$v_{esc}$ = escape speed, the speed required to escape the effects of the gravity of $m_1$ and coast to 'infinity'.

Therefore, as $r \Rightarrow \infty$, $v \Rightarrow 0$.

Parabolic orbits are difficult to achieve and rarely exist.

Hyperbolic Motion $(-\infty < a < 0,\ E > 0,\ e > 1)$

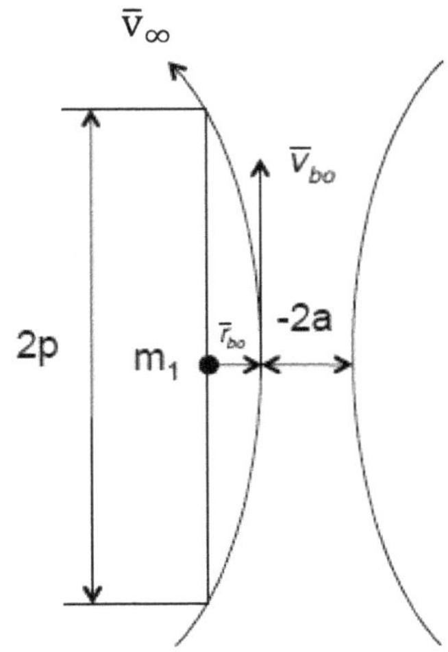


Mass $m_2$ must depart on a hyperbolic trajectory in order to have residual speed remaining after escaping the gravitational attraction of $m_1$.

- $v_\infty = |\bar{v}_\infty|$ = hyperbolic excess speed, the residual speed $m_2$ will have after escaping the gravitational attraction of $m_1$

- $v_{bo} = |\bar{v}_{bo}|$ = burnout speed, the speed of $m_2$ required at the burnout point, $r_{bo}$, to achieve the desired hyperbolic excess speed

The desired hyperbolic excess speed is normally known, and used to compute the required burnout speed from the equation

$$v_{bo} = \sqrt{v_\infty^2 + \frac{2\mu}{r_{bo}}}$$

Once $v_{bo}$ is determined, an appropriate velocity change, $\Delta v$, can be applied at the burnout point, $\bar{r}_{bo}$ to achieve the burnout speed. A comparison of orbital speeds on the various conic sections is provided in the series of figures below.

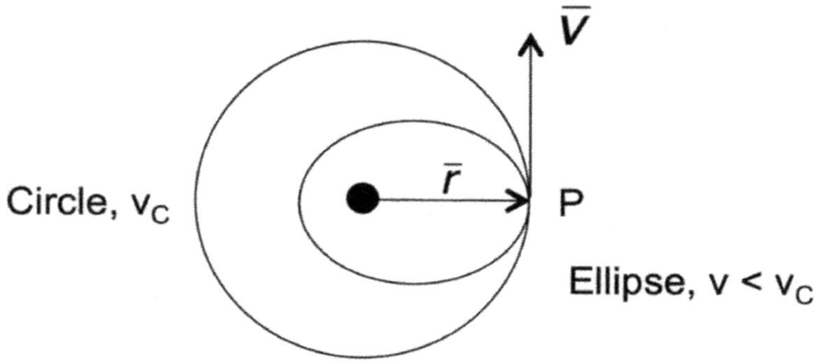
$\bar{V}$
$\bar{r}$
Circle, $v_C$
P
Ellipse, $v < v_C$

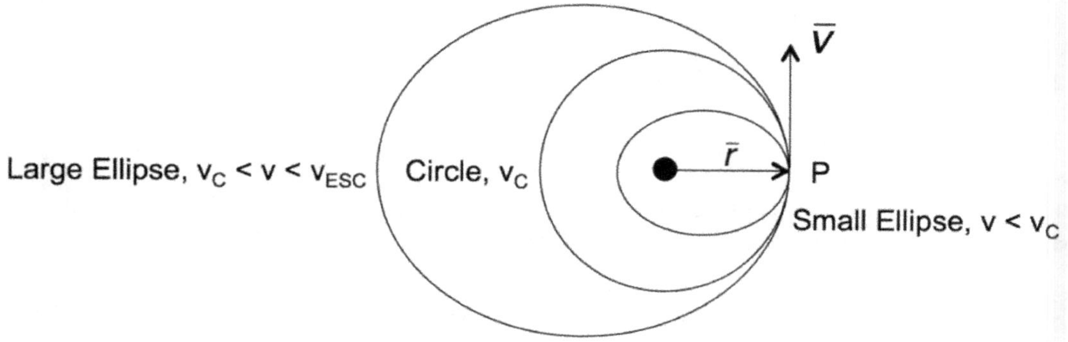
$\bar{V}$
$\bar{r}$
Large Ellipse, $v_C < v < v_{ESC}$
Circle, $v_C$
P
Small Ellipse, $v < v_C$

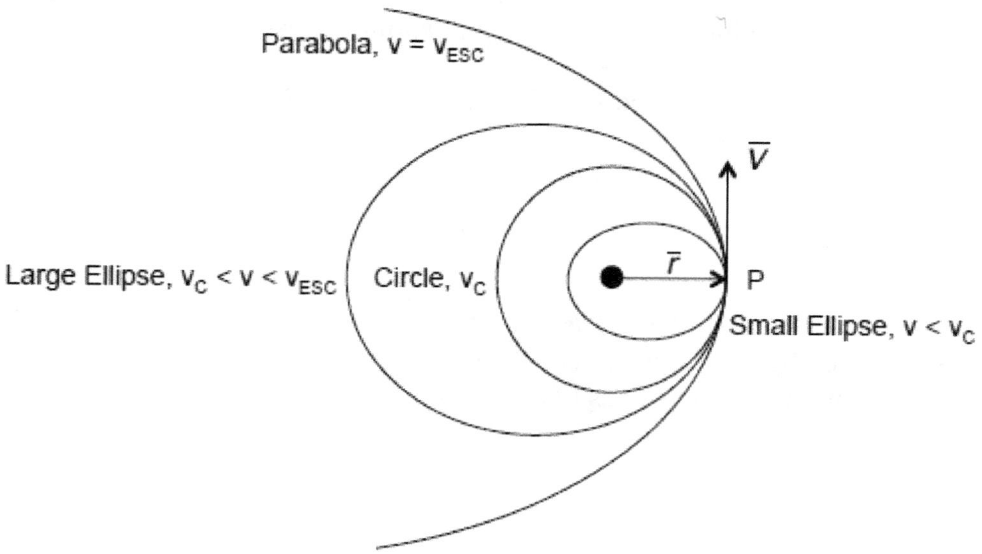
Parabola, $v = v_{ESC}$
$\bar{V}$
$\bar{r}$
Large Ellipse, $v_C < v < v_{ESC}$
Circle, $v_C$
P
Small Ellipse, $v < v_C$

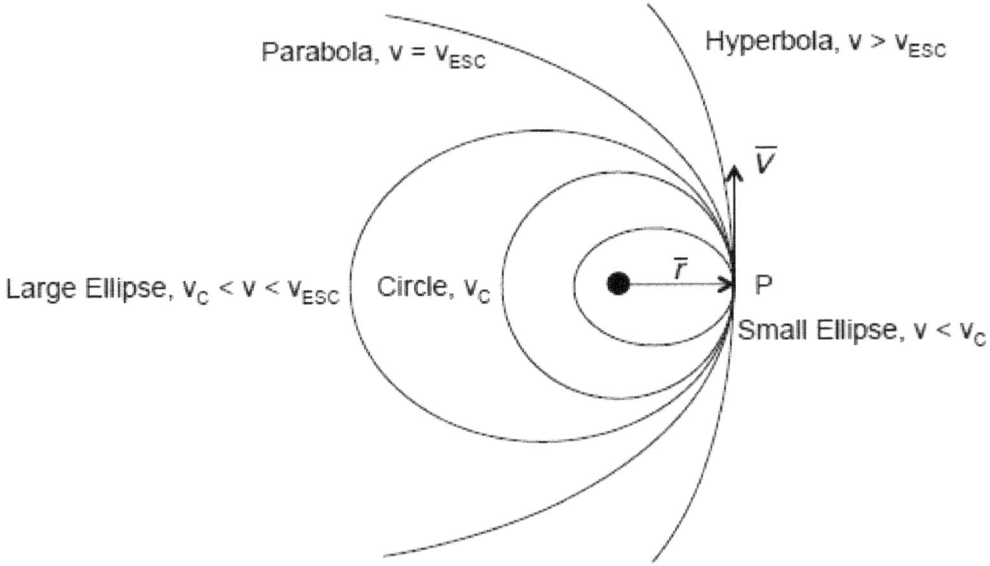


<u>Example 8.1</u>

For an Earth satellite in a circular orbit having a period of 100 minutes, calculate its altitude, circular orbit speed, and escape speed.

Solution:

$$T = 2\pi\sqrt{\frac{a^3}{\mu}} = 100 \text{ min} = 6{,}000 \text{ sec}$$

$$a = \sqrt[3]{\frac{\mu T^2}{4\pi^2}} = \sqrt[3]{\frac{3.986\times10^5(6{,}000)^2}{4\pi^2}} = 7{,}136.6 \text{ km} = r$$

$$(\text{alt}) = a - R_E = 7{,}136.6 - 6{,}378.1$$

$$\Rightarrow (\text{alt}) = 758.5 \text{ km}$$

$$v_c = \sqrt{\frac{\mu}{r}} = \sqrt{\frac{3.986\times10^5}{7{,}136.6}}$$

$$\Rightarrow v_c = 7.47 \frac{\text{km}}{\text{s}}$$

$$v_{esc} = \sqrt{2}v_c = (1.414)(7.47)$$

$$\Rightarrow v_{esc} = 10.57 \frac{\text{km}}{\text{s}}$$

Example 8.2

A starship is in a parking orbit about planet M, which has the following properties

$$\mu = 4.0 \times 10^6 \frac{km^3}{s^2} \qquad a = 12{,}000\ km \qquad e = 0.3$$

Find the speed of the starship at the point where $\nu = 60^o$.

Solution:

$$r = \frac{a(1-e^2)}{1+e\cos\nu} = \frac{12{,}000[1-(0.3)^2]}{1+(0.3)\cos 60^o} = 9.495.7\ km$$

$$v = \sqrt{\left[\mu\left(\frac{2}{r}-\frac{1}{a}\right)\right]} = \sqrt{\left[(4.0\times10^6)\left(\frac{2}{9{,}495.7}-\frac{1}{12{,}000}\right)\right]}$$

$$\Rightarrow v = 22.56\ \frac{km}{s}$$

Example 8.3

A satellite is in a circular Earth orbit at an altitude of 250 km. At a certain point the speed is increased by 45%. If that point becomes the periapsis of the new hyperbolic trajectory, find the semi-major axis and eccentricity of the new orbit, and the corresponding hyperbolic excess speed.

Solution:

$$r_p = R_E + (alt) = 6{,}378.1 + 250 = 6{,}628.1\ km$$

$$v_c = \sqrt{\frac{\mu}{r_p}} = \sqrt{\frac{3.986\times10^5}{6{,}628.1}} = 7.75\ \frac{km}{s}$$

$$v_{bo} = 1.45v_c = (1.45)(7.75) = 11.24\ \frac{km}{s}$$

$$v = \sqrt{\left[\mu\left(\frac{2}{r}-\frac{1}{a}\right)\right]}$$

$$a = \frac{1}{\frac{2}{r_p}-\frac{v_p^2}{\mu}} = \frac{1}{\frac{2}{6{,}628.1}-\frac{(11.24)^2}{3.986\times10^5}}$$

$$\Rightarrow \quad a = -65{,}756.0 \text{ km}$$

$$r_p = a(1 - e)$$

$$e = 1 - \frac{r_p}{a} = 1 - \frac{6{,}628.1}{-65{,}756.0}$$

$$\Rightarrow \quad e = 1.101$$

$$v_{bo} = \sqrt{v_\infty^2 + \frac{2\mu}{r_p}}$$

$$v_\infty = \sqrt{v_{bo}^2 - \frac{2\mu}{r_p}} = \sqrt{(11.24)^2 - \frac{2(3.986 \times 10^5)}{6{,}628.1}}$$

$$\Rightarrow \quad v_\infty = 2.46 \ \frac{\text{km}}{\text{s}}$$

Problems

8.1 Consider an Earth satellite in a circular orbit at an altitude of 300 km. Determine
(a) the speed of the satellite in its circular orbit
(b) the minimum increase in speed required to place this satellite on a trajectory to escape Earth's gravity
(c) the increase in speed required for the satellite to achieve a trajectory having a hyperbolic excess speed of $0.6 \frac{\text{km}}{\text{s}}$.
(Ans. (a) $v_c = 7.73 \ \frac{\text{km}}{\text{s}}$; (b) $\Delta v = 3.2 \frac{\text{km}}{\text{s}}$; (c) $\Delta v = 3.22 \frac{\text{km}}{\text{s}}$)

8.2 Assuming the mass of the moon to be $\frac{1}{81.28}$ times the mass of Earth, and that the period of the moon's orbit about Earth is 27.3217 days, calculate the eccentricity of the moon's orbit about Earth and its speed at periapsis.
(Ans. $e = 0.054, v_p = 1.08 \frac{\text{km}}{\text{s}}$)

8.3 For an Earth satellite in a circular orbit at an altitude of 600 km, find its resulting hyperbolic excess speed if the satellite's orbital speed is increased instantaneously by $4.0 \frac{\text{km}}{\text{s}}$.
(Ans. $v_\infty = 4.40 \ \frac{\text{km}}{\text{s}}$)

8.4 Consider a spacecraft orbiting Earth in an elliptical orbit having $a = 8{,}000$ km and $e = 0.1$. Compare the velocity change required to achieve Earth escape for the case where the spacecraft departs from perigee to the velocity change required if the spacecraft departed from apogee.
(Ans. $\Delta v_p = 2.18\ \frac{km}{s}, \Delta v_a = 3.60\ \frac{km}{s}$)

8.5 An Earth satellite is in a circular parking orbit at an altitude of 400 km. If it is planned for the satellite to depart Earth on a hyperbolic trajectory, calculate the velocity increase required if the satellite is to have a hyperbolic excess speed of $3.00\ \frac{km}{s}$.
(Ans. $\Delta v = 3.58\ \frac{km}{s}$)

8.6 Consider an Earth satellite in a circular orbit at an altitude of 500 km. Determine the $\Delta v$ required for the satellite to depart the circular orbit and enter an elliptical orbit which will cause the satellite to impact Earth's surface at a true anomaly $300^o$.
(Ans. $\Delta v = -0.193\ \frac{km}{s}$)

8.7 A spacecraft is orbiting Earth on a circular path at an altitude of 250 km. If its speed is increased by $0.2\ \frac{km}{s}$, find the eccentricity of the resulting elliptical orbit and its altitude at apoapsis.
(Ans. $e = 0.052$, $(alt)_a = 981.2$ km)

8.8 A satellite is placed in a circular polar orbit, i.e., passing over Earth's poles, at an altitude of 400 km. As the satellite passes over the North Pole, a retro-rocket is fired which produces a burst of negative thrust which instantaneously reduces its speed to a value which will ensure an equatorial landing. Calculate the required speed change to accomplish this mission.
(Ans. $\Delta v = -0.23\ \frac{km}{s}$)

# Module 9: Position and Velocity as a Function of Time

One more integral is needed to obtain enough information to formulate a complete solution to the Relative Two-Body Problem. The missing information is how a body moves as a function of time, which will be provided by Kepler's Equation. The development of Kepler's Equation begins with motion of $m_2$ about $m_1$ in the perifocal coordinate system defining the orbital plane.

## Motion in the Perifocal Coordinate System

Consider the four points at the intersections of the ellipse with the major and minor axes, points A, B, C, and D, as shown in the figure

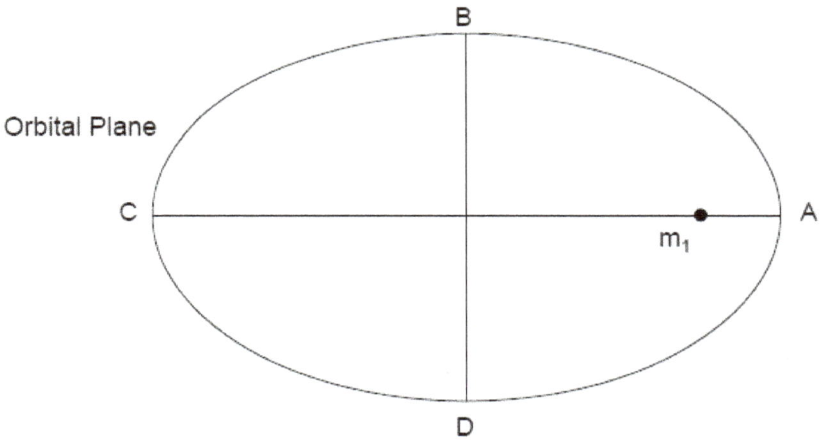


The orbital period, or the time it takes $m_2$ to travel one complete orbit, was determined from Kepler's Second Law, as

$$T = 2\pi\sqrt{\frac{a^3}{\mu}}$$

It follows that the time it takes to travel between the following points can be written in terms of the orbital period as

Travel from A to A: $\Delta t = T$
Travel from A to C: $\Delta t = 0.5T$
Travel from C to A: $\Delta t = 0.5T$

Since $m_2$ moves faster near periapsis and slower near apoapsis, the time it takes to travel from A to B, B to C, C to D, and D to A isn't simply 0.25T. More information about the dynamics of the problem is needed to determine its motion as a function of time.

The perifocal coordinate system is repeated here for reference during the development of the necessary relationships.

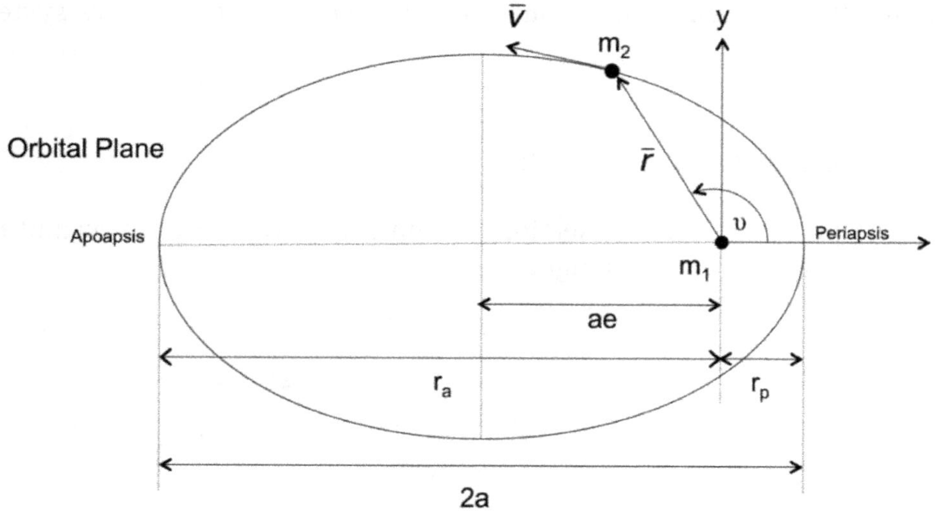


Kepler's Equation

The development of Kepler's Equation begins with circumscribing a circle about the perifocal coordinate system as follows

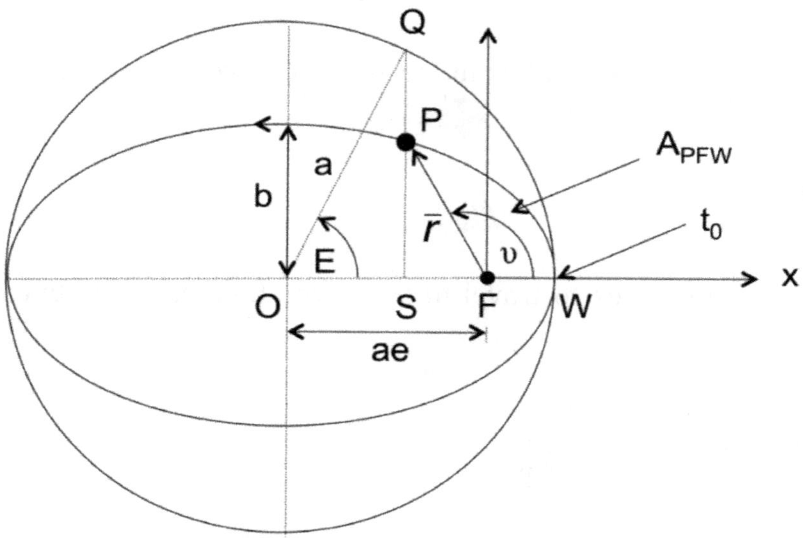

where

> $E$ = eccentric anomaly, the angle between periapsis and a line from the center of the circle circumscribed around the ellipse, to the intersection of the circle and a vertical line perpendicular to the major axis and passing through $m_2$

E is related to the true anomaly, ν, by the expression

$$\tan\left(\frac{\nu}{2}\right) = \sqrt{\frac{1+e}{1-e}}\tan\left(\frac{E}{2}\right) \tag{14}$$

Kepler's Second Law states that the radius vector from the Sun to any planet sweeps out equal areas in equal times.

From the figure above, area $A_{PFW}$ is the area swept out by the radius vector to $m_2$ in the time after periapsis passage, $\Delta t = t - t_0$. The total area of the ellipse is swept out during one orbital period, T.

Kepler's Second Law can be written in the form

$$\frac{\Delta t}{T} = \frac{A_{PFW}}{A_{Ellipse}}$$

Solving for the time after periapsis passage gives

$$\Delta t = \left(\frac{A_{PFW}}{\pi ab}\right) T$$

$A_{PFW}$ can be determined as follows

$$A_{PFW} = A_{PSW} - A_{PSF}$$

$$A_{PFW} = \left(\frac{b}{2}\right) A_{QSW} - \left(\frac{1}{2}\right)(ae - a\cos E)\left(\frac{b}{a}\right)(a\sin E)$$

$$A_{PFW} = \left(\frac{b}{a}\right)(A_{QOW} - A_{QOS}) - \left(\frac{ab}{2}\right)(e\sin E - \cos E\sin E)$$

$$A_{PFW} = \left(\frac{b}{a}\right)\left(\frac{a^2E}{2} - \frac{1}{2}(a\cos E)(a\cos E)\right) - \left(\frac{ab}{2}\right)(e\sin E - \cos E\sin E)$$

$$A_{PFW} = \left(\frac{b}{a}\right)\left(\frac{a^2}{2}\right)(E - \sin E\cos E) - \left(\frac{ab}{2}\right)(e\sin E - \cos E\sin E)$$

$$\Rightarrow A_{PFW} = \left(\frac{ab}{2}\right)(E - e\sin E)$$

Solving for $\Delta t$ gives

$$\Delta t = t - t_0 = \left[\frac{ab(E - e\sin E)}{2\pi ab}\right](2\pi)\sqrt{\frac{a^3}{\mu}}$$

$$\Rightarrow\ t - t_0 = \sqrt{\frac{a^3}{\mu}}(E - e\cos E)$$

Defining the mean anomaly, M, as the angular position of $m_2$ (measured from periapsis) as if it was moving with average angular velocity, n, called the 'mean motion', as

$$M = n(t - t_0)$$

where

$$n = \sqrt{\frac{\mu}{a^3}}$$

Therefore

$$M = n(t - t_0) = E - e\sin E \qquad (15)$$

Equation (15) is known as 'Kepler's Equation' and was also first presented in 1609 in *Astronomia Nova*. It can also be derived using dynamics instead of the geometrical derivation provided here. However, the geometrical derivation is more common and understandable. This form of Equation (15) is applicable to circular and elliptical orbits.

Kepler's Equation, along with Equation (14), represents the $12^{th}$ and final integral. Since all twelve required integrals have been found (six for the relative motion problem), the position and velocity of $m_2$ as a function of time can now be determined.

The important result of Kepler's Equation is that it relates the position of $m_2$ in its orbit to the time since periapsis passage. There are two types of problems that can be solved using Kepler's Equation. The first type of problem can be described

Type I: Calculate the time it takes for a spacecraft to travel from periapsis to a known point on its orbit, i.e., if $\nu$ is known, calculate E, then calculate $\Delta t = t - t_0$.

This type of application is more common and the solution is straight forward using Equations (14) and (15) above.

Example 9.1

An enemy ballistic missile is detected at the apogee point of its trajectory and found to have an altitude of 100 km and an eccentricity of 0.2. This missile must be intercepted during its descent before it reaches an altitude of 8 km. Determine the amount of time available to intercept this missile after detection.

Solution:

$$r_1 = R_E + (alt)_1 = 6{,}378.1 + 100 = 6{,}478.1 \text{ km}$$

$$r_2 = R_E + (alt)_2 = 6{,}378.1 + 8 = 6{,}386.1 \text{ km}$$

$$E_1 = 180^o = \pi \text{ rad}$$

$$a = \frac{r_1}{1+e} = \frac{6{,}478.1}{1+0.2} = 5{,}398.4 \text{ km}$$

$$r_2 = \frac{a(1-e^2)}{1+e\cos\nu_2}$$

$$\nu_2 = \cos^{-1}\left\{\frac{a(1-e^2)}{er_2} - \frac{1}{e}\right\} = \cos^{-1}\left\{\frac{5{,}398.4[1-(0.2)^2]}{0.2(6{,}386.1)} - \frac{1}{0.2}\right\} = 160.5^o$$

Since the intercept must take place after the apogee point of the missile's orbit

$$\Rightarrow \quad \nu_2 = 360^o - 160.5^o = 199.5^o$$

$$E_2 = 2\tan^{-1}\left\{\sqrt{\left[\frac{1-e}{1+e}\right]}\tan\left(\frac{\nu_2}{2}\right)\right\} = 2\tan^{-1}\left\{\sqrt{\left[\frac{1-0.2}{1+0.2}\right]}\tan\left(\frac{199.5^o}{2}\right)\right\} = -156.2^o$$

$$E_2 = -156.2^o + 360^o$$

$$\Rightarrow \quad E_2 = 203.8^o = 3.557 \text{ rad}$$

$$t_2 - t_o = \left(\frac{1}{n}\right)[E_2 - e\sin E_2] = \sqrt{\frac{a^3}{\mu}}[E_2 - e\sin E_2]$$

$$t_2 - t_o = \sqrt{\frac{(5{,}398.4)^3}{3.986\times10^5}}[3.557 - (0.2)\sin 203.8^o] = 2{,}285.37 \text{ sec}$$

$$t_1 - t_0 = \left(\frac{1}{n}\right)[E_1 - e \sin 1] = \sqrt{\frac{a^3}{\mu}}[E_1 - e \sin E_1]$$

$$t_1 - t_0 = \sqrt{\frac{(5{,}398.4)^3}{3.986 \times 10^5}}[\pi - (0.2)\sin 180^\circ] = 3.634 \text{ rad}$$

$$t_1 - t_0 = 1{,}973.69$$

$$\Delta t = (t_2 - t_0) - (t_1 - t_0) = (2{,}285.37) - (1{,}973.69)$$

$$\Rightarrow \quad \Delta t = 311.68 \text{ sec} = 5.195 \text{ min}$$

The second type of problem that can be solved using Kepler's Equation is described as

Type II: Calculate the true anomaly of $m_2$ at a given time after periapsis passage, i.e., if $\Delta t$ is known, calculate $\nu$.

This type of application is more difficult and computationally more intensive. Kepler's Equation is a transcendental equation, which means E cannot be solved for directly. E must be determined using a non-linear root finding technique, e.g., the Newton-Raphson Root Finding Method, an iterative solution method, which is discussed below.

Newton-Raphson Root Finding Method

Writing Equation (15) as

$$M = n(t - t_0) = E - e \sin E$$

An iterative solution for the eccentric anomaly can be expressed

$$E_{i+1} = E_i - \frac{M - E_i + e \cos E_i}{e \cos E_i - 1} \qquad i = 0, 1, 2 \ldots \tag{16}$$

where the initial value is set as

$$E_0 = M - n(t - t_0)$$

Iteration will proceed until convergence criteria are satisfied by

$$|E_{i+1} - E_i| < \delta$$

where $\delta$ is a specified tolerance $\sim 10^{-4}$ to $10^{-6}$.

For small values of e, Equation (16) will achieve quadratic convergence and generally within 3-4 iterations.

If e is large, the error in the solution could also be large.

Forms of Equation (16) can also be developed for both parabolic and hyperbolic orbits.

Example 9.2

A satellite is in an elliptical orbit where a = 7,400 km and e = 0.08. Use the Newton-Raphson Root Finding Method to calculate the eccentric anomaly at a time 50 minutes after perigee passage. Use four significant digits and perform two iterations only.

Solution:

$$\Delta t = 50(60) = 3{,}000 \text{ sec}$$

$$M = n\Delta t = \sqrt{\frac{\mu}{a^3}}\Delta t = \sqrt{\frac{3.986\times10^5}{(7{,}400)^3}}(3{,}000) = 2.9754 \text{ rad}$$

As a first guess of the eccentric anomaly let

$$E_0 = M = 2.9754 \text{ rad } (= 170.4767^o)$$

Then

$$E_{i+1} = E_i - \frac{M - E_i + e\sin E_i}{e\cos E_i - 1}, i = 0,1,2,\ldots,n$$

For i = 0:

$$E_1 = E_0 - \frac{M - E_0 + e\sin E_0}{e\cos E_0 - 1} = 2.9754 - \frac{2.9754 - 2.9754 + (0.08)\sin(170.4767^o)}{(0.08)\cos(170.4767^o) - 1}$$

$$E_1 = 2.9877 \ (= 171.1808^o)$$

For i = 1:

$$E_2 = E_1 - \frac{M - E_1 + e\sin E_1}{e\cos E_1 - 1} = 2.9877 - \frac{2.9754 - 2.9877 + (0.08)\sin(171.1808^o)}{(0.08)\cos(171.1808^o) - 1}$$

$$\Rightarrow \ E_2 = 2.9877 \ (= 171.1808^o)$$

Differenced Kepler's Equation

Kepler's Equation can be modified to calculate the $\Delta t$ between any two points on an orbit where a and e are known, instead of the time after periapsis passage. To do so, Kepler's Equation can be applied at both points to calculate the time-of-flight between them.

Consider the figure showing location of $m_2$ at time 1 and time 2.

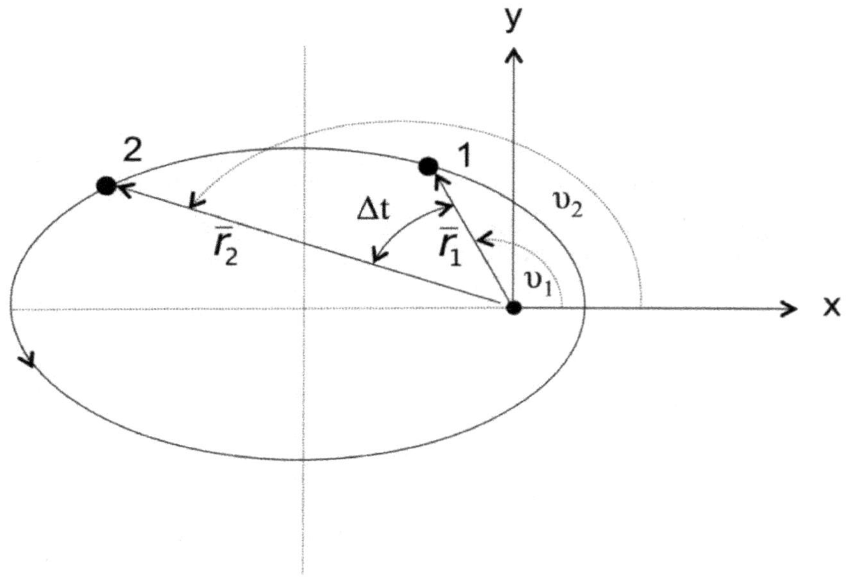


At position 1

$$n(t_1 - t_0) = E_1 - e \sin E_1$$

At position 2

$$n(t_2 - t_0) = E_2 - e \sin E_1$$

Subtracting Kepler's Equation for position 1 from that of position 2 gives the Differenced Kepler's Equation for elliptical orbits as

$$n(t_2 - t_1) = E_2 - E_1 - e\,(\sin E_2 - \sin E_1) \tag{17}$$

where

$$n = \sqrt{\frac{\mu}{a^3}}$$

$$E_1 = 2\tan^{-1}\left[\sqrt{\frac{1-e}{1+e}}\tan\left(\frac{\nu_1}{2}\right)\right]$$

$$E_2 = 2\tan^{-1}\left[\sqrt{\frac{1-e}{1+e}}\tan\left(\frac{\nu_2}{2}\right)\right]$$

Equation (17) is also applicable to circular and elliptical and orbits. Similar Differenced Kepler's Equations for parabolic and hyperbolic orbits can also be derived and are provided below.

Parabolic orbits:

$$t_2 - t_1 = \frac{1}{2}\sqrt{\frac{p^3}{\mu}}\left[\tan\left(\frac{\nu_2}{2}\right) + \frac{\tan^3\left(\frac{\nu_2}{2}\right)}{3} - \tan\left(\frac{\nu_1}{2}\right) - \frac{\tan^3\left(\frac{\nu_1}{2}\right)}{3}\right]$$

where

$$p = \frac{h^2}{\mu}$$

Hyperbolic orbits:

$$t_2 - t_1 = \sqrt{\frac{p^3}{\mu(e^2-1)^3}}\,(e\sinh F_2 - F_2 - e\sinh F_1 + F_2)$$

where $F$ is the hyperbolic eccentric anomaly and

$$F_1 = \ln\left[\frac{\sqrt{e+1}+\sqrt{e-1}\tan\left(\frac{\nu_1}{2}\right)}{\sqrt{e+1}+\sqrt{e-1}\tan\left(\frac{\nu_1}{2}\right)}\right]$$

$$F_2 = \ln\left[\frac{\sqrt{e+1}+\sqrt{e-1}\tan\left(\frac{\nu_2}{2}\right)}{\sqrt{e+1}+\sqrt{e-1}\tan\left(\frac{\nu_2}{2}\right)}\right]$$

Example 9.3

An Earth satellite has a semi-major axis of 10,550 km and an eccentricity of 0.15. At point A on the orbit the true anomaly is 42° and at point B on the orbit the true anomaly is 183°. Calculate the time it takes for the satellite to travel from point A to point B.

Solution:

$$E_A = 2\tan^{-1}\left\{\sqrt{\left[\frac{1-e}{1+e}\right]}\tan\left(\frac{\nu_A}{2}\right)\right\} = 2\tan^{-1}\left\{\sqrt{\left[\frac{1-0.15}{1+0.15}\right]}\tan\left(\frac{42^o}{2}\right)\right\}$$

$$E_A = 36.53^o = 0.6376 \text{ rad}$$

$$E_B = 2\tan^{-1}\left\{\sqrt{\left[\frac{1-e}{1+e}\right]}\tan\left(\frac{\nu_B}{2}\right)\right\} = 2\tan^{-1}\left\{\sqrt{\left[\frac{1-0.15}{1+0.15}\right]}\tan\left(\frac{183^o}{2}\right)\right\}$$

$$E_B = 183.49^o = 3.2025 \text{ rad}$$

$$t_B - t_A = \sqrt{\frac{a^3}{\mu}}\left[E_B - E_A - e(\sin E_B - \sin E_A)\right]$$

$$t_B - t_A = \sqrt{\frac{(10{,}550)^3}{3.986\times10^5}}\left[3.2025 - 0.6376 - 0.15(\sin 183.49^o - \sin 36.53^o)\right]$$

$$\Rightarrow \; t_B - t_A = 4{,}571.2 \text{ sec} = 76.19 \text{ min} = 1.27 \text{ hrs}$$

Position and Velocity Vectors in the Perifocal Coordinate System

Using the six integrals found for the Relative Two-Body Problem, the expressions for the position and velocity of $m_2$ in the perifocal coordinate system can be written in terms of orbital elements as

$$\bar{r} = \left[\frac{a(1-e^2)}{1+e\cos\nu}\right](\cos\nu\,\bar{\imath} + \sin\nu\,\bar{\jmath}) = a(\cos E - e)\,\bar{\imath} + \left(a\sqrt{1-e^2}\,\sin E\right)\bar{\jmath}$$

$$\bar{v} = \sqrt{\frac{\mu}{a(1-e^2)}}\left[-\sin\nu\,\bar{\imath} + (e+\cos\nu)\,\bar{\jmath}\right] = \frac{\sqrt{\mu a}}{r}\sin E\,\bar{\imath} + \frac{\sqrt{\mu a(1-e^2)}}{r}\cos E\,\bar{\jmath}$$

The Center-of-Mass Integrals can then be used to obtain the six vector components of position and velocity in the ECI coordinate system.

## Alternate Solutions to Kepler's Problem

The calculation of the position and velocity of a spacecraft in its orbit after a specified amount of time is known as Kepler's Problem. In other words, to find a spacecraft's future position and velocity given its position and velocity vectors at a specified time.
Kepler's Problem is also known as 'orbit propagation'.

The solution using the concepts of the Relative Two-Body Problem addressed above requires conversion back-and-forth between orbital elements and position and velocity vectors. Those algorithms are described in Module 10.

Two alternate solutions to Kepler's Problem, which have been historically include the f and g function solution, which is a closed-form solution, and the f and g series solution. These alternate solutions are briefly discussed below. Further details about these methods are available in Ref. [1].

## f and g Function Solutions

Consider a spacecraft's position and velocity are given at time $t_0$ as $\bar{r}_0$ and $\bar{v}_0$. The position and velocity at any future time can be written in terms of those known vectors as

$$\bar{r} = f\bar{r}_0 + g\,\bar{v}_0$$

$$\bar{v} = \dot{f}\bar{r}_0 + \dot{g}\,\bar{v}_0$$

Circular and Elliptical Orbits:

For these types of conic section motion f, g, $\dot{f}$, and $\dot{g}$ are defined as

$$f = 1 - \left(\frac{a}{r_0}\right)[1 - \cos(E - E_0]$$

$$g = (t - t_0) + \sqrt{\frac{a^3}{\mu}}[\sin(E - E_0) - (E - E_0)]$$

$$\dot{f} = \frac{-\sqrt{\mu a}}{r r_0}\sin\,(E - E_0)$$

$$\dot{g} = 1 - \left(\frac{a}{r}\right)[1 - \cos\,(E - E_{0)}]$$

These closed-form solutions are very useful in Preliminary/Initial Orbit Determination, POD/IOD.

If the true anomaly is known instead of the eccentric anomaly, the equations become

$$f = 1 - \left(\frac{a}{p}\right)[1 - \cos(\nu - \nu_0]$$

$$g = \frac{rr_0 \sin(\nu - \nu_0)}{\sqrt{h}}$$

$$\dot{f} = -\sqrt{\frac{\mu}{p}} \tan\left(\frac{\nu - \nu_0}{2}\right)\left\{\frac{[1 - \cos(\nu - \nu_0)]}{p} - \left(\frac{1}{r}\right) - \left(\frac{1}{r_0}\right)\right\}$$

$$\dot{g} = 1 - \left(\frac{r_0}{p}\right)[1 - \cos(\nu - \nu_{0)}]$$

Parabolic Orbits:

For a spacecraft on a parabolic orbit, if the parabolic anomaly, B, of the new position is known, the functions f, g, $\dot{f}$, and $\dot{g}$ are defined as

$$f = \frac{1 - B^2 + 2BB_0}{1 + B_0^2}$$

$$g = \frac{p^2(B - B_0)(1 + BB_0)}{2h}$$

$$\dot{f} = \frac{4h(B - B_0)}{p^2(1 + B^2)(1 + B_0^2)}$$

$$\dot{g} = \frac{1 - B_0^2 + 2BB_0}{1 + B^2}$$

Hyperbolic Orbits:

For hyperbolic orbits, if the hyperbolic anomaly, H, of the new position is known, the functions f, g, $\dot{f}$, and $\dot{g}$ become

$$f = 1 - \left(\frac{a}{r_0}\right)[1 - \cosh(H - H_0]$$

$$g = (t - t_0) + \sqrt{\frac{(-a)^3}{\mu}}[\sin h(H - H_0) - (H - H_0)]$$

$$\dot{f} = -\sqrt{\frac{\mu}{p}} \tanh\left(\frac{H - H_0}{2}\right)\left\{\frac{[1 - \cosh(H - H_0)]}{p} - \left(\frac{1}{r}\right) - \left(\frac{1}{r_0}\right)\right\}$$

$$\dot{g} = 1 - \left(\frac{r_0}{p}\right)[1 - \cosh(H - H_{0)}]$$

f and g Series Solutions

Consider again that a spacecraft's position and velocity are given at time $t_0$ as $\bar{r}_0$ and $\bar{v}_0$. The position and velocity at any future time can be written in terms of the known vectors, as

$$\bar{r} = f\,\bar{r}_0 + g\,\bar{v}_0$$

$$\bar{v} = \dot{f}\,\bar{r}_0 + \dot{g}\,\bar{v}_0$$

Here, the functions f, g, $\dot{f}$, and $\dot{g}$ are represented as infinite series in time, t. These functions, through seven (or eight) terms, provided by Ref. [1] are

$$f = 1 - \left(\frac{1}{2}\right)u_0 t^2 + \left(\frac{1}{2}\right)u_0 p_0 t^3 + \left(\frac{1}{24}\right)(3u_0 q_0 - 15u_0 p_0^2 + u_0^2)t^4$$

$$+ \left(\frac{1}{8}\right)(7u_0 p_0^3 - 3u_0 p_0 q_0 - u_0^2 p_0)t^5$$

$$+ \left(\frac{1}{720}\right)(630u_0 p_0^2 q_0 - 24u_0^2 q_0 - u_0^3 - 45u_0 q_0^2 - 945u_0 p_0^4 + 210u_0^2 p_0^2)t^6$$

$$+ \left(\frac{1}{5{,}040}\right)(882u_0^2 p_0 q_0 - 3{,}150u_0^2 p_0^3 - 9{,}450u_0 p_0^3 q_0 + 1{,}575u_0 p_0 q_0^2$$

$$+ 63u_0^3 p_0 - 10{,}395u_0 p_0^5)t^7$$

$$+ \left(\frac{1}{40{,}320}\right)(1{,}107u_0^2 q_0^2 - 24{,}570u_0^2 p_0^2 q_0 - 2{,}205u_0^3 p_0^2 + 51{,}975u_0^2 p_0^4$$

$$- 42{,}525u_0 p_0^2 q_0^2 + 155{,}925u_0 p_0^4 q_0 + 1{,}575u_0 q_0^3 + 117u_0^3 q_0$$

$$- 135{,}135u_0 p_0^6 + u_0^4)t^8$$

$$g = t - \left(\frac{1}{6}\right)u_0 t^3 + \left(\frac{1}{4}\right)u_0 p_0 t^4 + \left(\frac{1}{120}\right)(9u_0 q_0 - 45u_0 q_0^2 + u_0^2)t^5$$

$$+ \left(\frac{1}{360}\right)(210u_0 p_0^3 - 90u_0 p_0 q_0 - 15u_0^2 p_0)t^6$$

$$+ \left(\frac{1}{5{,}040}\right)(3{,}150u_0 p_0^2 q_0 - 54u_0^2 q_0 - 225u_0 q_0^2 - 4{,}725u_0 p_0^4 + 630u_0^2 p_0^2)t^7$$

$$+ \left(\frac{1}{40{,}320}\right)(3{,}024u_0^2 p_0 q_0 - 12{,}600u_0^2 p_0^3 - 56{,}700u_0 p_0^3 q_0 + 9{,}450u_0 p_0 q_0^2$$

$$+ 62{,}370u_0 p_0^5 + 126u_0^3 p_0)t^8$$

$$\dot{f} = -u_0 t + \left(\frac{3}{2}\right) u_0 p_0 t^2 + \left(\frac{1}{6}\right)(3u_0 q_0 - 15u_0 p_0^2 + u_0^2)t^3$$

$$+ \left(\frac{5}{8}\right)(7u_0 p_0^3 - 3u_0 p_0 q_0 - u_0^2 p_0)t^4$$

$$+ \left(\frac{1}{120}\right)(630u_0 p_0^2 q_0 - 24u_0^2 q_0 - u_0^3 - 45u_0 q_0^2 - 945u_0 p_0^4 + 210u_0^2 p_0^2)t^5$$

$$+ \left(\frac{1}{720}\right)(882u_0^2 p_0 q_0 - 3{,}150u_0^2 p_0^3 - 9{,}450u_0 p_0^3 q_0 + 1{,}575u_0 p_0 q_0^2$$

$$+ 63u_0^3 p_0 - 10{,}395u_0 p_0^5)t^6$$

$$+ \left(\frac{1}{5{,}040}\right)(1{,}107u_0^2 q_0^2 - 24{,}570u_0^2 p_0^2 q_0 - 2{,}205u_0^3 p_0^2 + 51{,}975u_0^2 p_0^4$$

$$- 42{,}525u_0 p_0^2 q_0^2 + 155{,}925u_0 p_0^4 q_0 + 1{,}575u_0 q_0^3 + 117u_0^3 q_0$$

$$- 135{,}135u_0 p_0^6 + u_0^4)t^7$$

$$g = 1 - \left(\frac{1}{2}\right) u_0 t^2 + u_0 p_0 t^3 + \left(\frac{1}{24}\right)(9u_0 q_0 - 45u_0 q_0^2 + u_0^2)t^4$$

$$+ \left(\frac{1}{60}\right)(210u_0 p_0^3 - 90u_0 p_0 q_0 - 15u_0^2 p_0)t^5$$

$$+ \left(\frac{1}{720}\right)(3{,}150u_0 p_0^2 q_0 - 54u_0^2 q_0 - 225u_0 q_0^2 - 4{,}725u_0 p_0^4 + 630u_0^2 p_0^2)t^6$$

$$+ \left(\frac{1}{5{,}040}\right)(3{,}024u_0^2 p_0 q_0 - 12{,}600u_0^2 p_0^3 - 56{,}700u_0 p_0^3 q_0 + 9{,}450u_0 p_0 q_0^2$$

$$+ 62{,}370u_0 p_0^5 + 126u_0^3 p_0)t^7$$

where

$$u_0 = \frac{\mu}{r_0^3}$$

$$p_0 = \frac{\bar{r}_0 \cdot \bar{v}_0}{r_0^2}$$

$$q_0 = \frac{v_0^2 - r_0^2 u_0}{r_0^2}$$

$t =$ time since periapsis passage

The solution accuracy will vary with the value of t, the eccentricity, and r. Generally, convergence is achieved for the conditions

$$|t| < \frac{(1-e)^3\left[-2e+\sqrt{4e^2+3e}\right]\left[(3+2e)-\sqrt{4e^2+3e}\right]^3}{108\left[e+\sqrt{4e^2+3e}\right]}$$

$$r = a(1-e)\left[1+\frac{e}{2(1-e)^3}t^2-\frac{e(1+3e)}{24(1-e)^6}t^4+\cdots\right]$$

Problems

9.1 A space-based interceptor is parked in a circular Earth orbit at an altitude of 500 km and must perform a fast intercept of a target located in another circular orbit at an altitude of 4,000 km. A rocket firing occurs placing the interceptor on a trajectory to hit the target at a point where the true anomaly of the intercept trajectory is 110°. Determine the time from rocket firing to target intercept.
(Ans. $\Delta t = 2{,}072.3$ sec $= 33.54$ min $= 0.576$ hrs)

9.2 An Earth satellite has a perigee altitude of 200 km and an apogee altitude of 900 km. Calculate the amount of time during each orbit that the satellite remains at an altitude above 400 km.
(Ans. $\Delta t = 62.7$ min)

9.3 A spacecraft has a perigee radius of 8,600 km and an apogee radius of 20,000 km. Find the time to fly from perigee to a true anomaly of 115°.
(Ans. $\Delta t = 54.23$ min)

9.4 For a satellite having a perigee radius of 7,000 km and an apogee radius of 10.000 km, calculate
(a) the difference in true anomaly between the times of 0.5 hr and 1.5 hr after perigee passage
(b) the total area that is swept out by the radius vector during the same time interval.
(Ans. (a) $\Delta \nu = 128.71°$; $\Delta A = 1.031 \times 10^8$ km$^2$)

9.5 A spacecraft has a period of 18.621 hours and a perigee radius of 14,297 km. Compute the radius and speed of the spacecraft at a time of 12 hours after perigee passage.
(Ans. $r = 53{,}548.2$ km, $v = 1.93\ \frac{km}{s}$)

9.6 Consider the case of an Earth satellite where $r_p = 7{,}500$ km and $r_a = 16{,}000$ km. Determine the value of the true anomaly 40 minutes after the satellite passes a true anomaly of 80°.
(Ans. $\nu = 142.03°$)

9.7 Calculate the time required to fly from apoapsis to the semi-minor axis in terms of the eccentricity, e, and orbital period, T, of an elliptical orbit.
(Ans. $\Delta t = T\left(\frac{1}{4} + \frac{e}{2\pi}\right)$)

9.8 If the eccentricity of an elliptical orbit is 0.25, find the time required to fly from a true anomaly of 90° to periapsis in terms of the orbital period, T.
(Ans. $\Delta t = 0.82873T$)

# Module 10: The Three-Dimensional Orbit in Space - ECI Vectors to Orbital Elements

The six integrals found for the Relative Two-Body Problem correspond to six quantities which are used to locate $m_2$ in three-dimensional space. These quantities, known as 'orbital elements', describe the size, shape, and orientation of the orbit of $m_2$, and its position within the orbital plane, relative to the ECI coordinate system. This orientation is commonly known as the three-dimensional orbit in space.

Likewise, the six orbital elements correspond to the six components of the position and velocity vectors of $m_2$ in its orbit, in both the ECI and perifocal coordinate systems.

Note that for this discussion, $\bar{r}$ and $\bar{v}$ will refer to the position and velocity of $m_2$ relative to $m_1$ in the perifocal coordinate system, while $\bar{R}$ and $\bar{V}$ will refer to the position and velocity of $m_2$ relative to $m_1$ in the ECI Coordinate System.

Except for the true anomaly, the orbital elements are <u>constant</u> under the assumptions of the Relative Two-Body Problem, while the position and velocity vectors change with time.

The Keplerian set of orbital elements is the classical and most common set, but other sets are also used. The Keplerian orbital elements are defined as follows:

$a$ = semi-major axis
$e$ = eccentricity
$i$ = inclination
$\Omega$ = longitude of the ascending node
$\omega$ = argument of periapsis
$\nu$ = true anomaly (not constant) ($t_p$, the time of perigee passage, is often substituted for $\nu$.)

The semi-major axis and the eccentricity of the orbit are defined in the perifocal coordinate system shown below, where the x-axis extends from the center of $m_1$, through periapsis.

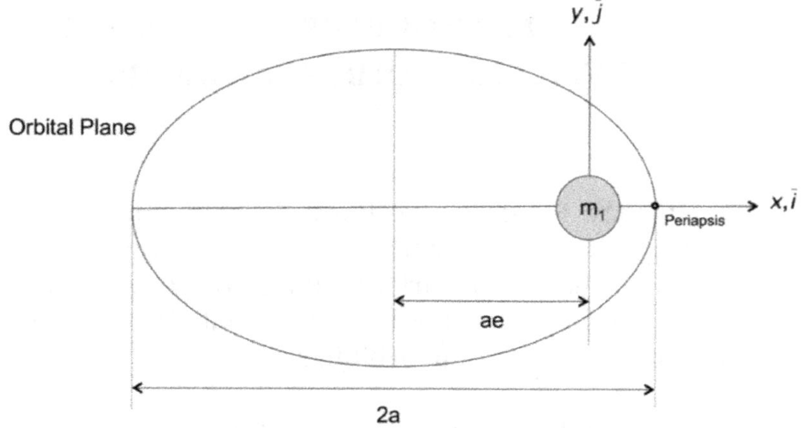


The remaining four orbital elements are defined by projecting the perifocal coordinate system, i.e., orbital plane, onto the ECI coordinate system at an angle of inclination i, thus creating a representation of a three-dimensional orbit-in-space. The other orbital elements are located on the following series of figures that define each element. Beginning with the ECI coordinate system, the XY-plane coincides with the equatorial plane of Earth.

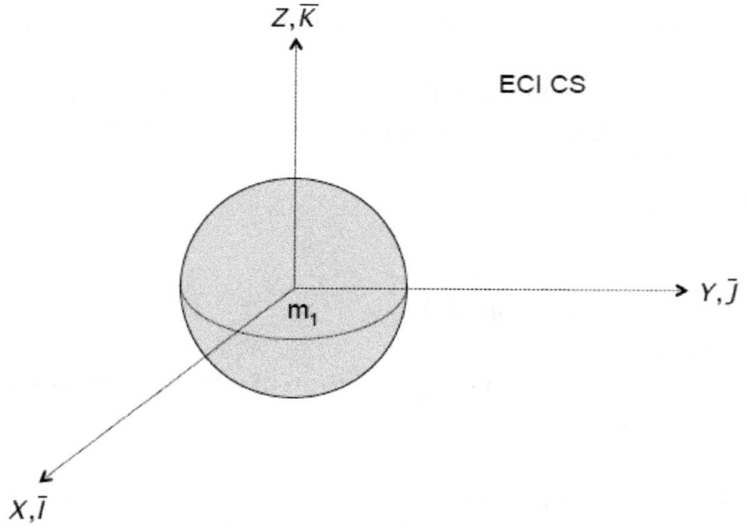


The orbital plane is 'sliced-through' Earth's equatorial plane at an orientation of angle i, which is the 'inclination' of the orbital plane relative to the equatorial plane.

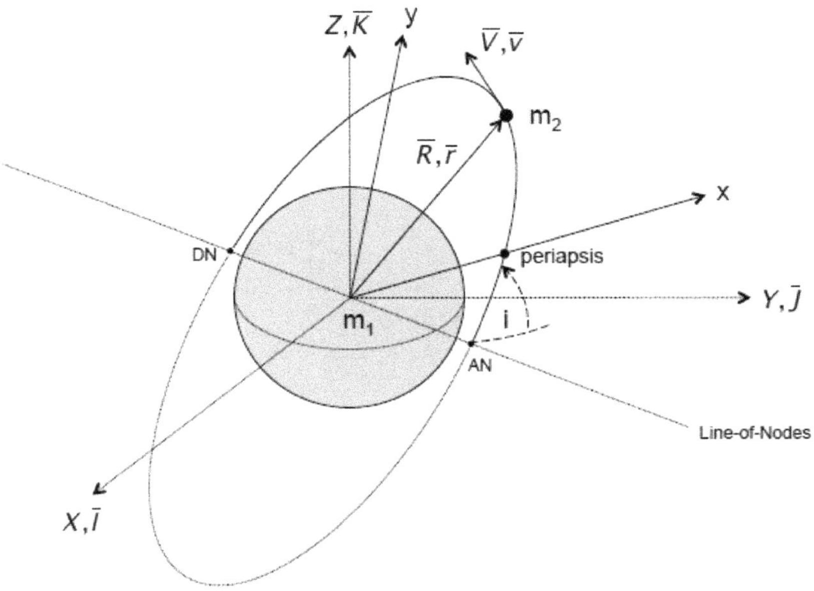


The intersection of the orbital plane with the equatorial plane defines the 'line-of-nodes,' which includes the two points on the orbit that pass through the equatorial plane. These points are the ascending node (AN) and the descending node (DN).

The point where $m_2$ passes from the Southern Hemisphere to the Northern Hemisphere is the AN, while the point where $m_2$ passes from the Northern Hemisphere to the Southern Hemisphere is the DN.

The inclination i, is also the angle between the ECI Z-axis and the angular momentum vector $\bar{h}$, which is perpendicular to the orbital plane and coincides with the z-axis of the perifocal coordinate system.

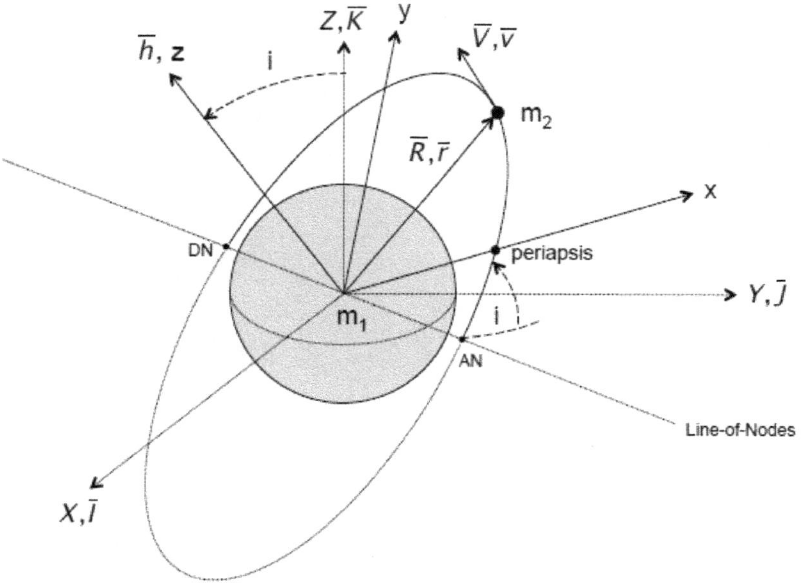


The inclination is measured CCW about the line-of-nodes (when looking through the AN toward $m_1$) from the equatorial plane to the orbital plane. It can have the values of $0^{o} \leq i \leq 180^{o}$.

The angle from the ECI X-axis to the AN, measured CCW about the ECI Z-axis, defines the 'longitude of the ascending node', $\Omega$, which can have the values $0^{o} \leq \Omega \leq 360^{o}$.

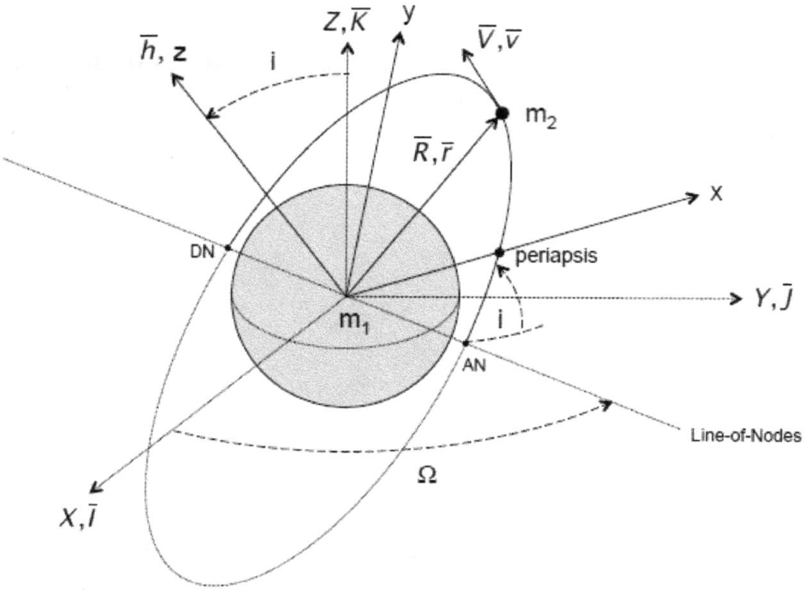

The angle from the line-of-nodes (at the AN), measured in in the orbital plane and in the direction of travel to periapsis (or perifocal x-axis), is the 'argument of periapsis', $\omega$, which can have values of $0^o \leq \omega \leq 360^o$.

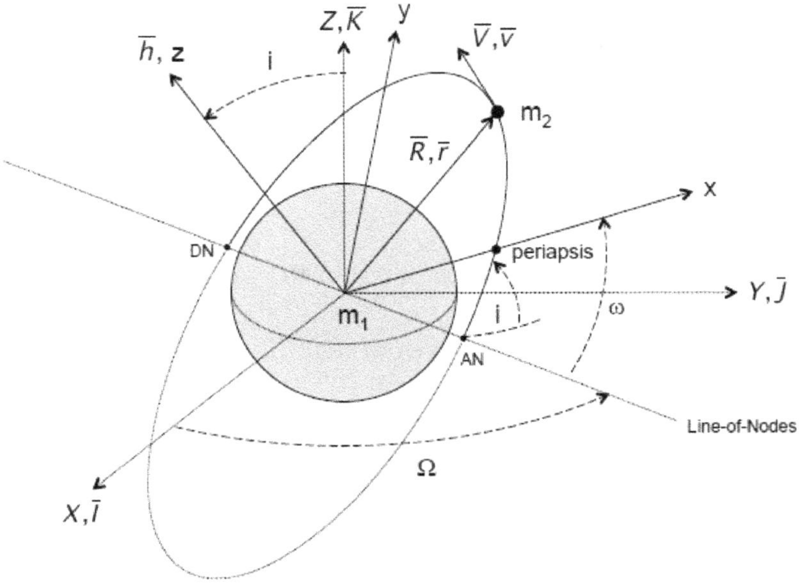


The final orbital element is the true anomaly, ν, which is the angle measured from periapsis to $\bar{r}$, in the orbital plane and in the direction of travel of $m_2$. It has the possible values $0^o \leq \nu \leq 360^o$.

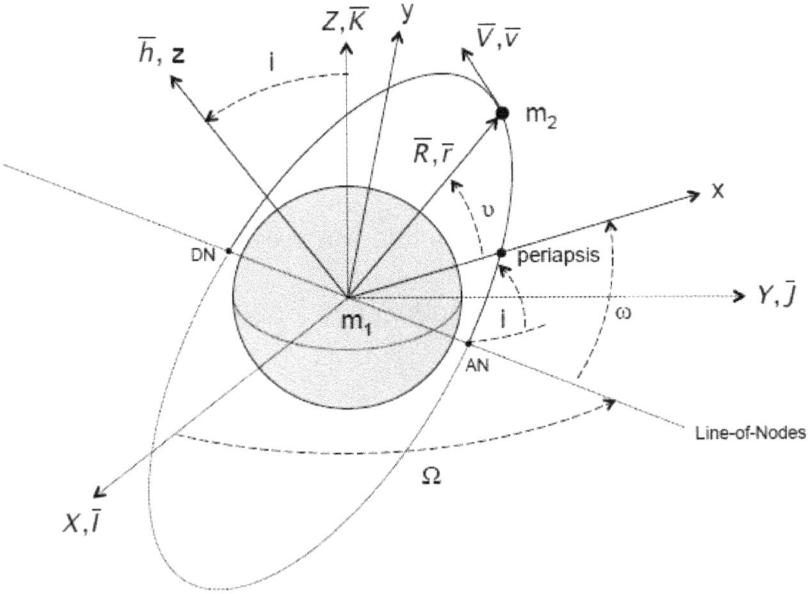

The Keplerian set of orbital elements $a$, $e$, $i$, $\Omega$, $\omega$, and $\nu$, completely define the location of $m_2$ in its orbit about $m_1$. In the figure above

- XYZ is the geocentric-equatorial (ECI) coordinate system.

- xyz is the perifocal coordinate system (x-axis directed through periapsis).

- $\bar{h}$ = angular momentum vector, normal to the orbital plane of $m_2$ about $m_1$.

- AN/DN = ascending/descending nodes.

- $\bar{r}$, $\bar{v}$ = position and velocity vectors of $m_2$ in the perifocal coordinate system.

- $\bar{R}$, $\bar{V}$ = position and velocity vectors of $m_2$ in ECI coordinate system.

- $a$, $e$, $i$, $\Omega$, $\omega$, $\nu$ = Keplerian orbital elements.

where

$a$ = semi-major axis, a measure of the size of the orbit.

$e$ = eccentricity, a measure of the shape of the orbit.

The angular orbital elements are defined as follows:

$i$ = inclination, the amount of tilt of the orbital plane relative to the ECI coordinate system, and the angle between the $\bar{K}$ and $\bar{h}$ vectors.

- The inclination is always measured counterclockwise (CCW) about the line-of-nodes, when looking at the ascending node, from the equatorial plane to the orbit plane.

- The possible values of the inclination are $0^o \leq i \leq 180^o$.

$\Omega$ = longitude of the ascending node, the angle between the $\bar{I}$ vector and the line-of-nodes at the ascending node.

- The longitude of the ascending node is always measured CCW about the Z-axis from the X-axis to the ascending node.

- The possible values of the longitude of the ascending node are $0^o \leq \Omega \leq 360^o$.

$\omega$ = argument of periapsis, the angle between the line-of-nodes and periapsis.

- The argument of periapsis is always measured about the z-axis in the direction of travel from the ascending node to periapsis.

- The possible values of the argument of periapsis are $0^o \leq \omega \leq 360^o$.

$\nu$ = true anomaly, the angle between periapsis and $\bar{r}$.

- The true anomaly is always measured about the z-axis in the direction of travel from periapsis to $\bar{r}$.

- The possible values of the true anomaly are $0^o \leq \nu \leq 360^o$.

Two other vectors are important relative to the orbit of $m_2$ about $m_1$, the node vector, $\bar{N}$, and the eccentricity vector, $\bar{e}$.

The node vector is defined as

$$\bar{N} = \bar{K} \times \bar{h} = \bar{K} \times (h_I \bar{I} + h_J \bar{J} + h_K \bar{K})$$

$$\Rightarrow \quad \bar{N} = h_J \bar{I} + h_I \bar{J}$$

Since $\bar{h}$ is constant, $\bar{N}$ will also be constant vector, in both magnitude and direction, pointing from $m_1$ along the line-of-nodes in the direction of the ascending node (AN).

The eccentricity vector is defined

$$\bar{e} = \left[\frac{V^2}{\mu} - \frac{1}{R}\right] \bar{R} - \frac{1}{\mu}[\bar{V} \cdot \bar{R}]\, \bar{V}$$

If $\bar{e}$ is evaluated at periapsis, where $\bar{R} = \bar{r}_p$ and $\bar{V} \cdot \bar{R} = 0$, then

$$\bar{e} = \left[\frac{V^2}{\mu} - \frac{1}{r_p}\right] \bar{r}_p$$

$$\bar{e} = \left[\frac{1+e}{a(1-e)} - \frac{1}{a(1-e)}\right] a(1-e)\, \bar{I}$$

$$\Rightarrow \quad \bar{e} = e\, \bar{I}$$

Since e is constant, $\bar{e}$ is also constant, and it will <u>always</u> point from $m_1$ towards periapsis with a magnitude equal to e.

The node vector and eccentricity vector are shown on the figure below.

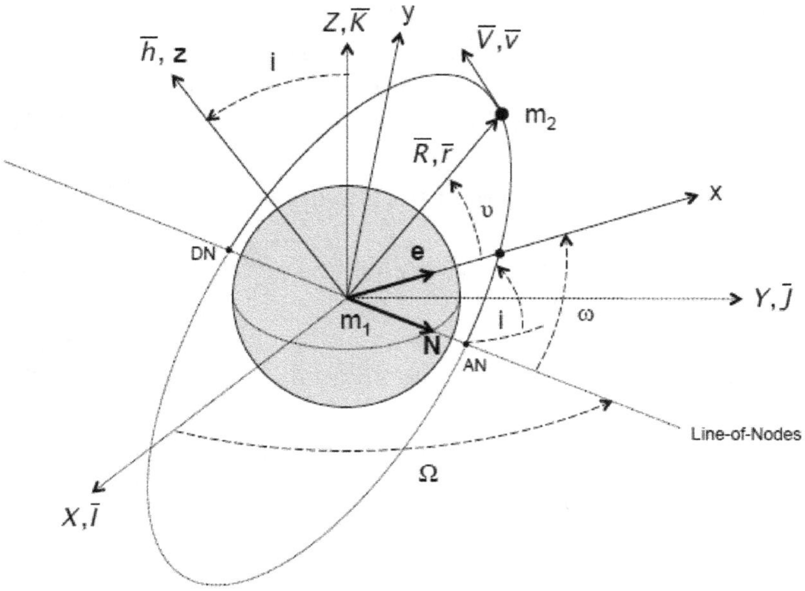


Two important tools in orbital mechanics are the calculation of the orbital elements from position and velocity vectors, and the calculation of position and velocity vectors from orbital elements. The algorithm for the conversion from ECI vectors to orbital elements is provided below, while the algorithm for the conversion from orbital elements to ECI vectors is presented in Module 11.

Tracking stations provide ECI position and velocity vectors which must be converted to the corresponding orbital elements. This problem can be formulated and the algorithm for the conversion is provided as

Given: $\bar{R}$ and $\bar{V}$ in the ECI coordinate system

Calculate: Keplerian set of orbital elements, a, e, i, Ω, ω, ν

1. Calculate e:

$$R = |\bar{R}|$$

$$V = |\bar{V}|$$

$$\bar{e} = \left[\frac{V^2}{\mu} - \frac{1}{R}\right]\bar{R} - \frac{1}{\mu}[\bar{R}\cdot\bar{V}]\,\bar{V}$$

$$\Rightarrow \quad e = |\bar{e}|$$

2. Calculate a:

$$\bar{h} = \bar{R}\times\bar{V} = h_I\,\bar{I} + h_J\,\bar{J} + h_K\,\bar{K}$$

$$h = |\bar{h}|$$

$$p = \frac{h^2}{\mu}$$

$$\Rightarrow \quad a = \frac{p}{1-e^2}$$

3. Calculate i:

$$\bar{h}\cdot\bar{K} = |\bar{h}||\bar{K}|\cos i$$

$$h_K = h(1)\cos i$$

$$\Rightarrow \quad i = \cos^{-1}\left(\frac{h_K}{h}\right) \quad \text{where } i \leq 180^o$$

4. Calculate $\Omega$:

$$\bar{N} = \bar{K} \times \bar{h} = -h_J\,\bar{I} + h_I\,\bar{J}$$

$$\bar{N} \cdot \bar{I} = |\bar{N}||\bar{I}| \cos \Omega$$

$$-h_J = N(1) \cos \Omega$$

$$\Omega = \cos^{-1}\left(\frac{-h_J}{N}\right)$$

Quadrant test: if $N_J > 0 \Rightarrow 0^o < \Omega < 180^o$
or, if $N_J < 0 \Rightarrow 180^o < \Omega < 360^o$

5. Calculate $\omega$:

$$\bar{N} \cdot \bar{e} = |\bar{N}||\bar{e}| \cos \omega = Ne \cos \omega$$

$$\omega = \cos^{-1}\left(\frac{\bar{N} \cdot \bar{e}}{Ne}\right)$$

Quadrant test: if $e_K > 0 \Rightarrow 0^o < \omega < 180^o$
or, if $e_K < 0 \Rightarrow 180^o < \omega < 360^o$

6. Calculate $\nu$:

$$\bar{e} \cdot \bar{R} = |\bar{e}||\bar{R}| \cos \nu = eR \cos \nu$$

$$\nu = \cos^{-1}\left(\frac{\bar{e} \cdot \bar{R}}{eR}\right)$$

Quadrant test: if $\bar{R} \cdot \bar{V} > 0 \Rightarrow 0^o < \nu < 180^o$
or, if $\bar{R} \cdot \bar{V} < 0 \Rightarrow 180^o < \nu < 360^o$

Example 10.1

Given the ECI position and velocity vectors of an Earth satellite below, calculate the corresponding set of Keplerian orbital elements.

$$\bar{R} = 6{,}500\,\bar{I} - 7{,}500\,\bar{J} - 2{,}500\,\bar{K}\ \text{km}$$

$$\bar{V} = 4\,\bar{I} + 3\,\bar{J} - 3\,\bar{K}\ \frac{\text{km}}{\text{s}}$$

Solution:

1. Calculate e:

$$R = |\bar{R}| = \sqrt{(6{,}500)^2 + (-7{,}500)^2 + (-2{,}500)^2} = 10{,}234.8 \text{ km}$$

$$V = |\bar{V}| = \sqrt{(4)^2 + (3)^2 + (-3)^2} = 5.83 \ \frac{\text{km}}{\text{s}}$$

$$\bar{e} = \left[\frac{V^2}{\mu} - \frac{1}{R}\right]\bar{R} - \frac{1}{\mu}[\bar{V}\cdot\bar{R}]\,\bar{V} = \left[\frac{(5.83)^2}{3.986\times 10^5} - \frac{1}{10{,}234.8}\right](6{,}500\,\bar{I} - 7{,}500\,\bar{J} - 2{,}500\,\bar{K})$$

$$-\frac{1}{3.986\times 10^5}[(4\,\bar{I} + 3\,\bar{J} - 3\,\bar{K})\cdot(6{,}500\,\bar{I} - 7{,}500\,\bar{J} - 2{,}500\,\bar{K})](4\,\bar{I} + 3\,\bar{J} - 3\,\bar{K})$$

$$\bar{e} = (-1.2436\times 10^{-5})(6{,}500\,\bar{I} - 7{,}500\,\bar{J} - 2{,}500\,\bar{K}) - \frac{11{,}000}{3.986\times 10^5}(4\,\bar{I} + 3\,\bar{J} - 3\,\bar{K})$$

$$\bar{e} = -0.191\,\bar{I} + 0.011\,\bar{J} + 0.114\,\bar{K}$$

$$e = |\bar{e}| = \sqrt{(-0.191)^2 + (0.011)^2 + (0.114)^2}$$

$$\Longrightarrow \quad e = 0.223$$

2. Calculate a:

$$\bar{h} = \bar{R}\times\bar{V} = (6{,}500\,\bar{I} - 7{,}500\,\bar{J} - 2{,}500\,\bar{K}) \times (4\,\bar{I} + 3\,\bar{J} - 3\,\bar{K})$$

$$\bar{h} = 30{,}000\,\bar{I} + 9{,}500\,\bar{J} + 49{,}500\,\bar{K}\ \frac{\text{km}^2}{\text{s}}$$

$$h = |\bar{h}| = \sqrt{(30{,}000)^2 + (9{,}500)^2 + (49{,}500)^2} = 58{,}655.8\ \frac{\text{km}^2}{\text{s}}$$

$$p = \frac{h^2}{\mu} = \frac{(58{,}655.8)^2}{3.986\times 10^5} = 8{,}631.5 \text{ km}$$

$$a = \frac{p}{1-e^2} = \frac{8{,}631.5}{1-(0.223)^2}$$

$$\Longrightarrow \quad a = 9{,}083.2 \text{ km}$$

3. Calculate i:

$$\bar{h} \cdot \bar{K} = |\bar{h}||\bar{K}| \cos i$$

$$h_K = h(1) \cos i$$

$$i = \cos^{-1}\left(\frac{h_K}{h}\right) = \cos^{-1}\left(\frac{49{,}500}{58{,}655.8}\right)$$

$$\Longrightarrow \quad i = 32.45^o$$

4. Calculate Ω:

$$\bar{N} = \bar{K} \times \bar{h} = \bar{K} \times (30{,}000\,\bar{I} + 9{,}500\,\bar{J} + 49{,}500\,\bar{K}) = -9{,}500\,\bar{I} + 30{,}000\,\bar{J}$$

$$N = |\bar{N}| = \sqrt{(-9{,}500)^2 + (30{,}000)^2} = 31{,}468.2$$

$$\bar{N} \cdot \bar{I} = |\bar{N}||\bar{I}| \cos \Omega$$

$$-h_J = N(1) \cos \Omega$$

$$\Omega = \cos^{-1}\left(\frac{-h_J}{N}\right) = \cos^{-1}\left(\frac{-9{,}500}{31{,}468.2}\right) = 107.57^o$$

Quadrant test:

Since $N_J = 30{,}000 > 0, \Longrightarrow 0^o < \Omega < 180^o$

$$\Longrightarrow \quad \Omega = 107.57^o$$

5. Calculate ω:

$$\bar{N} \cdot \bar{e} = (-9{,}500\,\bar{I} + 30{,}000\,\bar{J}) \cdot (-0.1912\,\bar{I} + 0.0105\,\bar{J} + 0.1139\,\bar{K}) = 2{,}130.91$$

$$\bar{N} \cdot \bar{e} = |\bar{N}||\bar{e}| \cos \omega = Ne \cos \omega$$

$$\omega = \cos^{-1}\left(\frac{\bar{N} \cdot \bar{e}}{Ne}\right) = \cos^{-1}\left[\frac{2{,}130.91}{(31{,}468.2)(0.223)}\right] = 72.3^o$$

Quadrant test:

Since $e_K = 0.1139 > 0 \Longrightarrow 0^o < \omega < 180^o$

$$\Longrightarrow \quad \omega = 72.3^o$$

6. Calculate ν:

$$\bar{e} \cdot \bar{R} = (-0.1912\,\bar{I} + 0.0105\,\bar{J} + 0.1139\,\bar{K}) \cdot (6{,}500\,\bar{I} - 7{,}500\,\bar{J} - 2{,}500\,\bar{K})$$

$$\bar{e} \cdot \bar{R} = -1{,}606.3$$

$$\bar{e} \cdot \bar{R} = |\bar{e}||\bar{R}| \cos\nu = eR\cos\nu$$

$$\nu = \cos^{-1}\left(\frac{\bar{e}\cdot\bar{R}}{eR}\right) = \cos^{-1}\left[\frac{-1{,}606.3}{(0.223)(10{,}234.8)}\right] = 134.7^{o}$$

Quadrant test:

$$\bar{R} \cdot \bar{V} = (6{,}500\,\bar{I} - 7{,}500\,\bar{J} - 2{,}500\,\bar{K}) \cdot (4\,\bar{I} + 3\,\bar{J} - 3\,\bar{K}) = 11{,}000$$

Since $\bar{R} \cdot \bar{V} = 11{,}000 > 0 \implies 0^{o} < \nu < 180^{o}$

$$\implies \nu = 134.7^{o}$$

Example 10.2

Calculate the corresponding set of Keplerian orbital elements from the ECI position and velocity vectors for an Earth satellite provided below.

$$\bar{R} = -2{,}290.826\,\bar{I} + 6{,}438.282\,\bar{J} + 5{,}179.268\,\bar{K}\ \text{km}$$

$$\bar{V} = -5.958\,\bar{I} - 3.258\,\bar{J} + 1.603\,\bar{K}\ \frac{\text{km}}{\text{s}}$$

Solution:

1. Calculate e:

$$R = |\bar{R}| = \sqrt{(-2{,}290.826)^2 + (6{,}438.282)^2 + (5{,}179.268)^2} = 8{,}574.62\ \text{km}$$

$$V = |\bar{V}| = \sqrt{(-5.958)^2 + (-3.258)^2 + (1.603)^2} = 6.977\ \frac{\text{km}}{\text{s}}$$

$$\bar{e} = \left[\frac{V^2}{\mu} - \frac{1}{R}\right]\bar{R} - \frac{1}{\mu}[\bar{V}\cdot\bar{R}]\,\bar{V} = \left[\frac{(6.977)^2}{3.986\times10^5} - \frac{1}{8{,}574.62}\right]\cdot$$

$$(-2{,}290.826\,\bar{I} + 6{,}438.282\,\bar{J} + 5{,}179.268\,\bar{K})$$

$$-\frac{1}{3.986\times10^5}[(-5.958\,\bar{I} - 3.258\,\bar{J} + 1.603\,\bar{K})\cdot$$

$$(-2{,}290.826\,\bar{I} + 6{,}438.282\,\bar{J} + 5{,}179.268\,\bar{K}](-5.958\,\bar{I} - 3.258\,\bar{J} + 1.603\,\bar{K})$$

$$\bar{e} = (5.501\times10^{-6})(-2{,}290.826\,\bar{I} + 6{,}438.282\,\bar{J} + 5{,}179.268\,\bar{K})$$

$$-\frac{975.971}{3.986\times10^5}(-5.958\,\bar{I} - 3.258\,\bar{J} + 1.603\,\bar{K})$$

$$\bar{e} = 0.0020\,\bar{I} + 0.0434\,\bar{J} + 0.0246\,\bar{K}$$

$$e = |\bar{e}| = \sqrt{(0.0020)^2 + (0.0434)^2 + (0.0246)^2}$$

$$\Longrightarrow \quad e = 0.050$$

2. Calculate a:

$$\bar{h} = \bar{R}\times\bar{V} = (-2{,}290.826\,\bar{I} + 6{,}438.282\,\bar{J} + 5{,}179.268\,\bar{K})$$

$$\times(-5.958\,\bar{I} - 3.258\,\bar{J} + 1.603\,\bar{K})$$

$$\bar{h} = 27{,}196.3\,\bar{I} - 27{,}185.9\,\bar{J} + 45{,}822.8\,\bar{K}\ \frac{km^2}{s}$$

$$h = |\bar{h}| = \sqrt{(27{,}196.3)^2 + (-27{,}185.9)^2 + (45{,}822.8)^2} = 59{,}820.07\ \frac{km^2}{s}$$

$$p = \frac{h^2}{\mu} = \frac{(59{,}820.07)^2}{3.986\times10^5} = 8{,}977.52\ km$$

$$a = \frac{p}{1-e^2} = \frac{8{,}977.52}{1-(0.050)^2}$$

$$\Longrightarrow \quad a = 9{,}000.0\ km$$

3. Calculate i:

$$\bar{h}\cdot\bar{K} = |\bar{h}||\bar{K}|\cos i$$

$$h_K = h(1)\cos i$$

$$i = \cos^{-1}\left(\frac{h_K}{h}\right) = \cos^{-1}\left(\frac{45{,}822.8}{59{,}820.07}\right)$$

$$\Rightarrow \quad i = 40.0^o$$

4. Calculate $\Omega$:

$$\bar{N} = \bar{K} \times \bar{h} = \bar{K} \times (27{,}196.3\,\bar{I} - 27{,}185.9\,\bar{J} + 45{,}822.8\,\bar{K})$$

$$\bar{N} = 27{,}185.9\,\bar{I} + 27{,}196.3\,\bar{J}$$

$$N = |\bar{N}| = \sqrt{(27{,}185.9)^2 + (27{,}196.3)^2} = 38{,}454.0$$

$$\bar{N} \cdot \bar{I} = |\bar{N}||\bar{I}| \cos\Omega$$

$$-h_J = N(1)\cos\Omega$$

$$\Omega = \cos^{-1}\left(\frac{-h_J}{N}\right) = \cos^{-1}\left(\frac{27{,}185.9}{38{,}454.0}\right) = 45.0^o$$

Quadrant test:

Since $N_J = 27{,}196.3 > 0, \Rightarrow 0^o < \Omega < 180^o$

$$\Rightarrow \quad \Omega = 45.1^o$$

5. Calculate $\omega$:

$$\bar{N} \cdot \bar{e} = (27{,}185.9\,\bar{I} + 27{,}196.3\,\bar{J}) \cdot (0.0020\,\bar{I} + 0.0434\,\bar{J} + 0.0246\,\bar{K})$$

$$\bar{N} \cdot \bar{e} = 1{,}234.69$$

$$\bar{N} \cdot \bar{e} = |\bar{N}||\bar{e}| \cos\omega = Ne\cos\omega$$

$$\omega = \cos^{-1}\left(\frac{\bar{N}\cdot\bar{e}}{Ne}\right) = \cos^{-1}\left[\frac{1{,}234.69}{(38{,}454.0)(0.050)}\right] = 50.0^o$$

Quadrant test:

Since $e_K = 0.0246 > 0 \Rightarrow 0^o < \omega < 180^o$

$$\Rightarrow \quad \omega = 50.0^o$$

6. Calculate ν:

$$\bar{e} \cdot \bar{R} = (0.0020\,\bar{I} + 0.0434\,\bar{J} + 0.0246\,\bar{K}) \cdot (-2{,}290.826\,\bar{I} + 6{,}438.282\,\bar{J}$$
$$+ 5{,}179.268\,\bar{K})$$

$$\bar{e} \cdot \bar{R} = 402.25$$

$$\bar{e} \cdot \bar{R} = |\bar{e}||\bar{R}| \cos \nu = eR \cos \nu$$

$$\nu = \cos^{-1}\left(\frac{\bar{e} \cdot \bar{R}}{eR}\right) = \cos^{-1}\left[\frac{402.25}{(0.050)(8{,}574.62)}\right] = 20.2^{o}$$

Quadrant test:

$$\bar{R} \cdot \bar{V} = (-2{,}290.826\,\bar{I} + 6{,}438.282\,\bar{J} + 5{,}179.268\,) \cdot$$
$$(-5.958\,\bar{I} - 3.258\,\bar{J} + 1.603\,\bar{K}) = 975.971$$

Since $\bar{R} \cdot \bar{V} = 975.971 > 0 \implies 0^{o} < \nu < 180^{o}$

$\implies \nu = 20.2^{o}$

Example 10.3

The ECI position and velocity vectors for an Earth satellite are provided below. Calculate the corresponding set of Keplerian orbital elements.

$$\bar{R} = 7{,}015.975\,\bar{I} + 6{,}355.011\,\bar{J} + 1{,}068.013\,\bar{K}\text{ km}$$

$$\bar{V} = 4.276\,\bar{I} - 4.512\,\bar{J} - 1.623\,\bar{K}\,\frac{\text{km}}{\text{s}}$$

Solution:

1. Calculate e:

$$R = |\bar{R}| = \sqrt{(7{,}015.975)^2 + (6{,}355.011)^2 + (1{,}068.013)^2} = 9{,}526.32\text{ km}$$

$$V = |\bar{V}| = \sqrt{(4.276)^2 + (-4.512)^2 + (-1.623)^2} = 6.425\,\frac{\text{km}}{\text{s}}$$

$$\bar{e} = \left[\frac{V^2}{\mu} - \frac{1}{R}\right]\bar{R} - \frac{1}{\mu}[\bar{V}\cdot\bar{R}]\,\bar{V} = \left[\frac{(6.425)^2}{3.986\times10^5} - \frac{1}{9{,}526.32}\right]\cdot$$

$$(7{,}015.975\,\bar{I} + 6{,}355.011\,\bar{J} + 1{,}068.013\,\bar{K})$$

$$-\frac{1}{3.986\times10^5}[(4.276\,\bar{I} - 4.512\,\bar{J} - 1.623\,\bar{K})\cdot$$

$$(7{,}015.975\,\bar{I} + 6{,}355.011\,\bar{J} + 1{,}068.013\,\bar{K})]\cdot(4.276\,\bar{I} - 4.512\,\bar{J} - 1.623\,\bar{K})$$

$$\bar{e} = (-1.4083\times10^{-5})(7{,}015.975\,\bar{I} + 6{,}355.011\,\bar{J} + 1{,}068.013\,\bar{K})$$

$$-\frac{-406.886}{3.986\times10^5}(4.276\,\bar{I} - 4.512\,\bar{J} - 1.623\,\bar{K})$$

$$\bar{e} = -0.0056\,\bar{I} - 0.0136\,\bar{J} - 0.0032\,\bar{K}$$

$$e = |\bar{e}| = \sqrt{(-0.0056)^2 + (-0.0136) + (-0.0032)^2}$$

$$\Longrightarrow\quad e = 0.015$$

2. Calculate a:

$$\bar{h} = \bar{R}\times\bar{V} = (7{,}015.975\,\bar{I} + 6{,}355.011\,\bar{J} + 1{,}068.013\,\bar{K})\times$$

$$(4.276\,\bar{I} - 4.512\,\bar{J} - 1.623\,\bar{K})$$

$$\bar{h} = -5{,}495.30\,\bar{I} + 15{,}953.75\,\bar{J} - 58{,}830.06\,\bar{K}\,\frac{km^2}{s}$$

$$h = |\bar{h}| = \sqrt{(-5{,}495.30)^2 + (15{,}953.75)^2 + (-58.830.06)^2} = 61{,}202.09\,\frac{km^2}{s}$$

$$p = \frac{h^2}{\mu} = \frac{(61{,}202.09)^2}{3.986\times10^5} = 9{,}397.13\ km$$

$$a = \frac{p}{1-e^2} = \frac{9{,}397.13}{1-(0.015)^2}$$

$$\Longrightarrow\quad a = 9{,}400.0\ km$$

3. Calculate i:

$$\bar{h} \cdot \bar{K} = |\bar{h}||\bar{K}| \cos i$$

$$h_K = h(1) \cos i$$

$$i = \cos^{-1}\left(\frac{h_K}{h}\right) = \cos^{-1}\left(\frac{-58{,}830.06}{61{,}202.09}\right)$$

$$\Rightarrow \quad i = 164.0^{o}$$

4. Calculate Ω:

$$\bar{N} = \bar{K} \times \bar{h} = \bar{K} \times (-5{,}495.30\,\bar{I} + 15{,}953.75\,\bar{J} - 58{,}830.06\,\bar{K})$$

$$= -15{,}953.75\,\bar{I} - 5{,}495.30\,\bar{J}$$

$$N = |\bar{N}| = \sqrt{(-15{,}953.75)^2 + (-5{,}495.30)^2} = 16{,}873.67$$

$$\bar{N} \cdot \bar{I} = |\bar{N}||\bar{I}| \cos \Omega$$

$$-h_J = N(1) \cos \Omega$$

$$\Omega = \cos^{-1}\left(\frac{-h_J}{N}\right) = \cos^{-1}\left(\frac{-15{,}953.75}{16{,}873.67}\right) = 161.0^{o}$$

Quadrant test:

Since $N_J = -5{,}495.30 < 0, \Rightarrow \; 180^{o} < \Omega < 360^{o}$

$$\Rightarrow \quad \Omega = 360^{o} - 161.0^{o} = 199.0^{o}$$

5. Calculate ω:

$$\bar{N} \cdot \bar{e} = (-15{,}953.75\,\bar{I} - 5{,}495.30\,\bar{J}) \cdot (-0.0056\,\bar{I} - 0.0136\,\bar{J} - 0.0032\,\bar{K})$$

$$\bar{N} \cdot \bar{e} = 164.05$$

$$\bar{N} \cdot \bar{e} = |\bar{N}||\bar{e}| \cos \omega = Ne \cos \omega$$

$$\omega = \cos^{-1}\left(\frac{\bar{N} \cdot \bar{e}}{Ne}\right) = \cos^{-1}\left[\frac{164.05}{(16{,}873.67)(0.015)}\right] = 49.6^{o}$$

Quadrant test:

Since $e_K = -0.00032 < 0 \Rightarrow 180^o < \omega < 360^o$

$\Rightarrow \omega = 360.0^o - 49.6^o = 310.4^o$

5. Calculate ν:

$\bar{e} \cdot \bar{R} = (-0.0056\,\bar{I} - 0.0136\,\bar{J} - 0.0032) \cdot$

$(7{,}015.975\,\bar{I} + 6{,}355.011\,\bar{J} + 1{,}068.013\,\bar{K})$

$\bar{e} \cdot \bar{R} = -129.18$

$\bar{e} \cdot \bar{R} = |\bar{e}||\bar{R}| \cos\nu = eR\cos\nu$

$$\nu = \cos^{-1}\left(\frac{\bar{e}\cdot\bar{R}}{eR}\right) = \cos^{-1}\left[\frac{-129.18}{(0.015)(9{,}526.32)}\right] = 154.7^o$$

Quadrant test:

$\bar{R} \cdot \bar{V} = (7{,}015.975\,\bar{I} + 6{,}355.011\,\bar{J} + 1{,}068.013\,\bar{K}) \cdot$

$(4.276\,\bar{I} - 4.512\,\bar{J} - 1.623\,\bar{K}) = -406.886$

Since $\bar{R} \cdot \bar{V} = -406.886 < 0 \Rightarrow 180^o < \nu < 360^o$

$\Rightarrow \nu = 360^o - 154.7^o = 205.4^o$

Classification of Orbits

Orbits are classified according to inclination, orbital period, and altitude, as

- $i < 90^o \Rightarrow$ orbit is a direct orbit, i.e., west-to-east
- $i = 90^o \Rightarrow$ orbit is a polar orbit, i.e., passes over the North and South Poles.
- $i > 90^o \Rightarrow$ orbit is a retrograde orbit, i.e., east-to-west
- $i = 0^o \Rightarrow$ orbit is an equatorial (direct) orbit
- $i = 180^o \Rightarrow$ orbit is an equatorial (retrograde) orbit

Geosynchronous orbits have a period equal to the rotation rate of Earth, which is one sidereal day (23 hr, 56 min, 4.09 sec), and the altitude is 35,800 km (r = 42,178 km) for a circular orbit.

Geostationary (GEO) orbits are geosynchronous orbits having $i = 0^o$.

Low-Earth Orbit (LEO): altitude ~ 300–1,500 km

Medium-Earth Orbit (MEO): altitude ~1,500–35,800 km

Highly Elliptical Orbit (HEO):

perigee altitude < 1,000 km; apogee altitude ~15,000–40,000 km

Problems

10.1 The position, velocity, eccentricity, and angular momentum vectors of an Earth satellite in the ECI coordinate system are given below. Calculate the Keplerian set of orbital elements for the satellite.

$$\bar{R} = 518..3\,\bar{I} - 29{,}695.5\,\bar{J}\ \text{km} \qquad \bar{V} = 1.64\,\bar{I} + 0.29\,\bar{J}\ \frac{\text{km}}{\text{s}}$$

$$\bar{e} = 0.02\,\bar{I} + 0.80\,\bar{J} \qquad \bar{h} = 48{,}850.93\,\bar{K}\ \frac{\text{km}^2}{\text{s}}$$

(Ans. $e = 0.8$, $a = 16{,}630.5$ km, $i = 0.0^o$, $\Omega =$ undefined, $\omega =$ undefined)

10.2 An Earth satellite has the ECI vectors and orbital parameters given below. Compute the Keplerian set of orbital elements for this satellite.

$$\bar{R} = 2{,}189.9\,\bar{I} + 240.7\,\bar{J} - 8{,}203.5\,\bar{K}\ \text{km} \qquad R = 8{,}494.2\ \text{km}$$

$$\bar{V} = -7.27\,\bar{I} - 1.31\,\bar{J} - 1.72\,\bar{K}\ \frac{\text{km}}{\text{s}} \qquad V = 7.58\,\frac{\text{km}}{\text{s}}$$

$$\bar{h} = -11{,}160.6\,\bar{I} + 63{,}406.1\,\bar{J} - 1{,}118.9\,\bar{K} \qquad h = 64{,}390.6\,\frac{\text{km}^2}{\text{s}}$$

$$\bar{N} = -63{,}406.1\,\bar{I} - 11{,}160.6\,\bar{J}\ \frac{\text{km}^2}{\text{s}} \qquad N = 64{,}380.8\,\frac{\text{km}^2}{\text{s}}$$

$$\bar{e} = 0.0191\,\bar{I} - 0.0006\,\bar{J} - 0.2259\,\bar{K} \qquad \bar{e}\cdot\bar{R} = 1{,}894.9\ \text{km}$$

$$\bar{N}\cdot\bar{e} = -1{,}204.4\,\frac{\text{km}^2}{\text{s}} \qquad \bar{R}\cdot\bar{V} = -2{,}125.9\ \frac{\text{km}^2}{\text{s}}$$

(Ans. $e = 0.227$, $a = 10{,}966.9$ km, $i = 91.0^o$, $\Omega = 190.0^o$, $\omega = 265.3^o$, $\nu = 349.3^o$)

10.3 Calculate the Keplerian orbital elements for an Earth satellite having the following ECI position and velocity vectors

$\overline{R} = 2{,}500\,\overline{I} + 16{,}000\,\overline{J} + 4{,}000\,\overline{K}\text{ km}$ $\qquad\qquad$ $\overline{V} = -3\,\overline{I} - \overline{J} + 5\,\overline{K}\,\frac{\text{km}}{\text{s}}$

(Ans. $e = 0.466$, $a = 31{,}163.1$ km, $i = 62.53^{o}$, $\Omega = 73.74^{o}$, $\omega = 22.10^{o}$, $\nu = 353.52^{o}$)

10.4 For the ECI position and velocity vectors for an Earth satellite given below, determine the corresponding Keplerian orbital elements.

$\overline{R} = -13{,}000\,\overline{K}\text{ km}$ $\qquad\qquad$ $\overline{V} = 4\,\overline{I} - 5\,\overline{J} + 6\,\overline{K}\,\frac{\text{km}}{\text{s}}$

(Ans. $e = 1.297$, $a = -25.433.6$ km, $i = 90.00^{o}$, $\Omega = 51.34^{o}$, $\omega = 344.95^{o}$, $\nu = 285.05^{o}$)

10.5 Find the Keplerian orbital elements of a satellite having the ECI position and velocity vectors provided below:

$\overline{R} = -1{,}020.988\,\overline{I} + 4{,}110.971\,\overline{J} + 5{,}391.578\,\overline{K}\text{ km}$

$\overline{V} = -5.716\,\overline{I} - 4.502\,\overline{J} + 2.572\,\overline{K}\,\frac{\text{km}}{\text{s}}$

(Ans. $a = 8{,}000$ km, $e = 0.100$, $i = 60.0^{o}$, $\Omega = 50.0^{o}$, $\omega = 40.0^{o}$, $\nu = 30.0^{o}$)

10.6 Calculate the Keplerian orbital elements of a satellite having the ECI position and velocity vectors

$\overline{R} = 8{,}948.767\,\overline{I} - 3{,}214.340\,\overline{J} - 663.275\,\overline{K}\text{ km}$

$\overline{V} = -2.275\,\overline{I} - 5.992\,\overline{J} - 0.774\,\overline{K}\,\frac{\text{km}}{\text{s}}$

(Ans. $a = 9{,}500$ km, $e = 0.010$, $i = 172.0^{o}$, $\Omega = 190.0^{o}$, $\omega = 320.0^{o}$, $\nu = 250.0^{o}$)

10.7 Consider the Earth satellite having the ECI position and velocity vectors of

$\overline{R} = -5{,}000\,\overline{I} - 8{,}000\,\overline{J} - 2{,}100\,\overline{K}\text{ km}$ $\qquad$ $\overline{V} = -4.0\,\overline{I} + 3.5\,\overline{J} - 3.0\,\overline{K}\,\frac{\text{km}}{\text{s}}$

Determine the Keplerian orbital elements for this spacecraft.

(Ans. $a = 8{,}812$ km, $e = 0.101$, $i = 147.09^{o}$, $\Omega = 78.11^{o}$, $\omega = 8.62^{o}$, $\nu = 194.92^{o}$)

10.8 Calculate the Keplerian orbital elements for the low-Earth satellite having the ECI position and velocity vectors

$$\bar{R} = -3{,}020.779\,\bar{I} + 5{,}232.142\,\bar{J} + 2{,}198.947\,\bar{K}\ \text{km}$$

$$\bar{V} = -7.067\,\bar{I} - 3.758\,\bar{J} + 0.101\,\bar{K}\ \frac{\text{km}}{\text{s}}$$

(Ans. $a = 6{,}652.56$ km, $e = 0.05$, $i = 20.0°$, $\Omega = 30.0°$, $\omega = 40.0°$, $\nu = 50.0°$)

# Module 11: The Three-Dimensional Orbit in Space - Orbital Elements to ECI Vectors

Another important tool in orbital mechanics is the calculation of the position and velocity vectors in the ECI coordinate system from the orbital elements. This calculation requires a coordinate transformation. To understand this process, it is easier first transform ECI coordinates to perifocal coordinates and then reverse the transformation. This can begin by viewing the relationship between the two coordinate systems in the figure below.

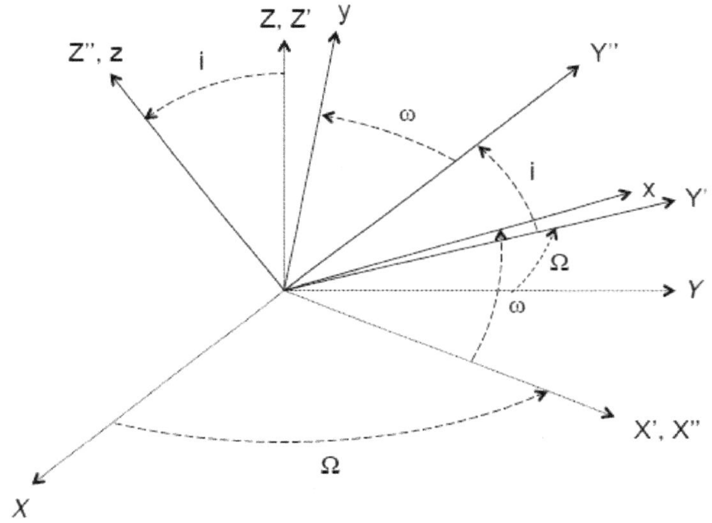


A series of three angular rotations are required to align the ECI coordinates with perifocal coordinates. The rotations, in the proper order, are

- Rotate XYZ (ECI) CCW about the Z-axis, through angle Ω, to obtain X′Y′Z′, where

$$\begin{bmatrix} X' \\ Y' \\ Z' \end{bmatrix} = T(\Omega) \begin{bmatrix} X \\ Y \\ Z \end{bmatrix} \quad , \quad T(\Omega) = \begin{bmatrix} \cos\Omega & -\sin\Omega & 0 \\ \sin\Omega & \cos\Omega & 0 \\ 0 & 0 & 1 \end{bmatrix}$$

Rotate X′Y′Z′ CCW about the X′-axis, through angle i, to obtain X″Y″Z″, where

$$\begin{bmatrix} X'' \\ Y'' \\ Z'' \end{bmatrix} = T(i) \begin{bmatrix} X' \\ Y' \\ Z' \end{bmatrix} \quad , \quad T(i) = \begin{bmatrix} 1 & 0 & 0 \\ 0 & \cos i & -\sin i \\ 0 & \sin i & \cos i \end{bmatrix}$$

- Rotate X″Y″Z″ CCW about the X″-axis, through angle ω, to obtain xyz (perifocal), where

$$\begin{bmatrix} x \\ y \\ z \end{bmatrix} = T(\omega) \begin{bmatrix} X''' \\ Y''' \\ Z''' \end{bmatrix} \quad , \quad T(\omega) = \begin{bmatrix} \cos\omega & -\sin\omega & 0 \\ \sin\omega & \omega & 0 \\ 0 & 0 & 1 \end{bmatrix}$$

- Combining these three transformations shows

$$\begin{bmatrix} x \\ y \\ z \end{bmatrix} = T(\omega)\, T(i)\, T(\Omega) \begin{bmatrix} X \\ Y \\ Z \end{bmatrix}$$

- Reversing these three transformations shows

$$\begin{bmatrix} X \\ Y \\ Z \end{bmatrix} = T^{-1}(\Omega)\, T^{-1}(i)\, T^{-1}(\omega) \begin{bmatrix} x \\ y \\ z \end{bmatrix} = [T] \begin{bmatrix} x \\ y \\ z \end{bmatrix}$$

where

$$[T] = T^{-1}(\Omega)\, T^{-1}(I)\, T^{-1}(\omega)$$

$$[T] = \begin{bmatrix} \cos\Omega & -\sin\Omega & 0 \\ \sin\Omega & \cos\Omega & 0 \\ 0 & 0 & 1 \end{bmatrix}^{-1} \begin{bmatrix} 1 & 0 & 0 \\ 0 & \cos i & -\sin i \\ 0 & \sin i & \cos i \end{bmatrix}^{-1} \begin{bmatrix} \cos\omega & -\sin\omega & 0 \\ \sin\omega & \omega & 0 \\ 0 & 0 & 1 \end{bmatrix}^{-1}$$

Since these transformations are orthogonal, their inverses are equal to their transforms, i.e., $[T]^{-1} = [T]^T$. Therefore

$$[T] = \begin{bmatrix} \cos\Omega & \sin\Omega & 0 \\ -\sin\Omega & \cos\Omega & 0 \\ 0 & 0 & 1 \end{bmatrix} \begin{bmatrix} 1 & 0 & 0 \\ 0 & \cos i & \sin i \\ 0 & -\sin i & \cos i \end{bmatrix} \begin{bmatrix} \cos\omega & \sin\omega & 0 \\ -\sin\omega & \omega & 0 \\ 0 & 0 & 1 \end{bmatrix}$$

Letting $c = \cos$, $s = \sin$, and multiplying gives the result

$$[T] = \begin{bmatrix} c\Omega\, c\omega - s\Omega\, c i s\,\omega & -c\Omega s\,\omega - s\Omega\, c i c\,\omega & s\Omega\, s i \\ s\Omega\, c\omega + c\Omega\, c i\, s\omega & -s\Omega s\,\omega + c\Omega\, c i\, c\omega & -c\Omega\, s i \\ s i\, s\omega & s i\, c\omega & c i \end{bmatrix}$$

The conversion problem can now be formulated and the algorithm for the calculations can be provided as follows.

Given: Keplerian set of orbital elements, a, e, i, Ω, ω, ν

Calculate: $\bar{R}$ and $\bar{V}$ In the ECI coordinate system

1. Calculate r:

$$p = a(1 - e^2)$$

$$\Rightarrow \quad r = \frac{p}{1+e\cos\nu}$$

2. Calculate [T]:

Let $c = \cos$, $s = \sin$, such that

$$\Rightarrow \quad [T] = \begin{bmatrix} c\Omega\, c\omega - s\Omega\, ci\, s\omega & -c\Omega s\,\omega - s\Omega\, ci\, c\omega & s\Omega\, si \\ s\Omega\, c\omega + c\Omega\, ci\, s\omega & -s\Omega\, s\omega + c\Omega\, ci\, c\omega & -c\Omega\, si \\ si\, s\omega & si\, c\omega & ci \end{bmatrix}$$

3. Calculate the perifocal position vector, $\bar{r}$, and the ECI position vector, $\bar{R}$:

$$\bar{r} = r\cos\nu\,\bar{I} + r\sin\nu\,\bar{J}$$

$$\Rightarrow \quad \bar{R} = \begin{bmatrix} R_I \\ R_J \\ R_K \end{bmatrix} = [T]\bar{r} = [T]\begin{bmatrix} r\cos\nu \\ r\sin\nu \\ 0 \end{bmatrix} = R_I\,\bar{I} + R_J\,\bar{J} + R_K\,\bar{K}$$

4. Calculate the perifocal velocity vector, $\bar{v}$, and the ECI velocity vector, $\bar{V}$:

$$\bar{v} = -\sqrt{\frac{\mu}{p}}\sin\nu\,\bar{I} + \sqrt{\frac{\mu}{p}}(e + \cos\nu)\,\bar{J}$$

$$\Rightarrow \quad \bar{V} = \begin{bmatrix} V_I \\ V_J \\ V_K \end{bmatrix} = [T]\bar{v} = [T]\begin{bmatrix} -\sqrt{\frac{\mu}{p}}\sin\nu \\ \sqrt{\frac{\mu}{p}}(e + \cos\nu) \\ 0 \end{bmatrix} = V_I\,\bar{I} + V_J\,\bar{J} + V_K\,\bar{K}$$

Example 11.1

A satellite has an apogee altitude of 4,121.9 km and the following orbital elements

$$a = 10{,}000 \text{ km} \qquad i = 30^o \qquad \Omega = 0^o \qquad \omega = 90^o \qquad E = 120^o$$

Calculate the position and velocity vectors of the satellite in the perifocal coordinate system.

Solution:

$$r_a = R_E + (alt)_a = 6{,}378.1 + 4{,}121.9 = 10{,}500 \text{ km}$$

$$e = \frac{r_a}{a} - 1 = \frac{10{,}500}{10{,}000} - 1 = 0.05$$

$$p = a(1 - e^2) = 10{,}000[1 - (0.05)^2] = 9{,}975 \text{ km}$$

$$\nu = 2\tan^{-1}\left[\sqrt{\frac{1+e}{1-e}}\tan\left(\frac{E}{2}\right)\right] = 2\tan^{-1}\left[\sqrt{\frac{1+0.05}{1-0.05}}\tan\left(\frac{120^o}{2}\right)\right] = 122.45^o$$

$$r = \frac{a(1-e^2)}{1+e\cos\nu} = \frac{10{,}000[1-(0.05)^2]}{1+(0.05)\cos 122.45^o} = 10{,}250 \text{ km}$$

$$\bar{r} = r\cos\nu\,\bar{\imath} + r\sin\nu\,\bar{\jmath} = 10{,}250\cos 122.45^o\,\bar{\imath} + 10{,}250\sin 122.45^o\,\bar{\jmath}$$

$$\Rightarrow \bar{r} = -5{,}500.0\,\bar{\imath} + 8{,}650.0\,\bar{\jmath} \text{ km}$$

$$\bar{v} = -\sqrt{\frac{\mu}{p}}\sin\nu\,\bar{\imath} + \sqrt{\frac{\mu}{p}}(e + \cos\nu)\,\bar{\jmath} = -\sqrt{\frac{3.986\times 10^5}{9{,}975}}\sin 122.45^o\,\bar{\imath}$$

$$+\sqrt{\frac{3.986\times 10^5}{9{,}975}}(0.05 + \cos 122.45^o)\,\bar{\jmath}$$

$$\Rightarrow \bar{v} = -5.33\,\bar{\imath} - 3.08\,\bar{\jmath}\,\frac{km}{s}$$

Example 11.2

Using the results of Example 11.1, calculate the ECI position and velocity vectors.

Solution:

$$[T] = \begin{bmatrix} c\Omega\, c\omega - s\Omega\, ci\, s\omega & -c\Omega s\,\omega - s\Omega\, ci\, c\omega & s\Omega\, si \\ s\Omega\, c\omega + c\Omega\, ci\, s\omega & -s\Omega\, s\omega + c\Omega\, ci\, c\omega & -c\Omega\, si \\ si\, s\omega & si\, c\omega & ci \end{bmatrix}$$

$$[T] = \begin{bmatrix} 0 & -1 & 0 \\ 0.866 & 0 & -0.5 \\ 0.5 & 0 & 0.866 \end{bmatrix}$$

$$\bar{R} = \begin{bmatrix} X \\ Y \\ Z \end{bmatrix} = [T]\bar{r} = \begin{bmatrix} 0 & -1 & 0 \\ 0.866 & 0 & -0.5 \\ 0.5 & 0 & 0.866 \end{bmatrix} \begin{bmatrix} -5{,}500.0 \\ 8{,}650.0 \\ 0 \end{bmatrix} = \begin{bmatrix} -8{,}650.0 \\ -4{,}763.0 \\ -2{,}750.0 \end{bmatrix}$$

$$\Rightarrow \; \bar{R} = -8{,}650\,\bar{I} - 4{,}763.0\,\bar{J} - 2{,}750.0\,\bar{K}\text{ km}$$

$$\bar{V} = \begin{bmatrix} X \\ Y \\ Z \end{bmatrix} = [T]\bar{v} = \begin{bmatrix} 0 & -1 & 0 \\ 0.866 & 0 & -0.5 \\ 0.5 & 0 & 0.866 \end{bmatrix} \begin{bmatrix} -5.33 \\ -3.08 \\ 0 \end{bmatrix} = \begin{bmatrix} -3.08 \\ -4.62 \\ -2.67 \end{bmatrix}$$

$$\Rightarrow \; \bar{V} = -3.08\,\bar{I} - 4.62\,\bar{J} - 2.67\,\bar{K}\,\frac{\text{km}}{\text{s}}$$

Example 11.3

An Earth satellite the following orbital elements

$a = 13{,}500$ km     $e = 0.100$     $i = 100^o$     $\Omega = 260^o$     $\omega = 275^o$     $\nu = 350^o$

Calculate the position and velocity vectors of the satellite in the ECI coordinate system.

Solution

$$p = a(1 - e^2) = 13{,}500[1 - (0.100)^2] = 13{,}365\text{ km}$$

$$r = \frac{a(1-e^2)}{1+e\cos\nu} = \frac{13{,}500[1-(0.100)^2]}{1+(0.100)\cos 350^o} = 12{,}166.8\text{ km}$$

$$\bar{r} = r\cos\nu\,\bar{\imath} + r\sin\nu\,\bar{\jmath} = 12{,}166.8\cos 350^o\,\bar{\imath} + 12{,}166.8\sin 350^o\,\bar{\jmath}$$

$$\Rightarrow \quad \bar{r} = 11{,}982.0\,\bar{\imath} - 2{,}112.7\,\bar{\jmath}\text{ km}$$

$$\bar{v} = -\sqrt{\frac{\mu}{p}}\sin\nu\,\bar{\imath} + \sqrt{\frac{\mu}{p}}\,(e+\cos\nu)\,\bar{\jmath} = -\sqrt{\frac{3.986\times10^5}{13{,}365}}\sin 350^\circ\,\bar{\imath}$$

$$+\sqrt{\frac{3.986\times10^5}{13{,}365}}\,(0.100+\cos 350^\circ)\,\bar{\jmath}$$

$$\Rightarrow \quad \bar{v} = -0.948\,\bar{\imath} + 5.924\,\bar{\jmath}\,\frac{\text{km}}{\text{s}}$$

$$[T] = \begin{bmatrix} c\Omega\, c\omega - s\Omega\, ci\, s\omega & -c\Omega s\,\omega - s\Omega\, ci\, c\omega & s\Omega\, si \\ s\Omega\, c\omega + c\Omega\, ci\, s\omega & -s\Omega\, s\omega + c\Omega\, ci\, c\omega & -c\Omega\, si \\ si\, s\omega & si\, c\omega & ci \end{bmatrix}$$

$$[T] = \begin{bmatrix} 0.1552 & -0.1879 & -0.9698 \\ -0.1159 & -0.9784 & 0.1710 \\ -0.9811 & 0.0858 & -0.1736 \end{bmatrix}$$

$$\bar{R} = \begin{bmatrix} X \\ Y \\ Z \end{bmatrix} = [T]\bar{r} = \begin{bmatrix} 0.1552 & -0.1879 & -0.9698 \\ -0.1159 & -0.9784 & 0.1710 \\ -0.9811 & 0.0858 & -0.1736 \end{bmatrix}\begin{bmatrix} 11{,}982.0 \\ -2112.7 \\ 0 \end{bmatrix} = \begin{bmatrix} 2{,}256.58 \\ 678.35 \\ -11{,}936.81 \end{bmatrix}$$

$$\Rightarrow \quad \bar{R} = 2{,}256.58\,\bar{I} + 678.35\,\bar{J} - 11{,}936.81\,\bar{K}\text{ km}$$

$$\bar{V} = \begin{bmatrix} X \\ Y \\ Z \end{bmatrix} = [T]\bar{v} = \begin{bmatrix} 0.1552 & -0.1879 & -0.9698 \\ -0.1159 & -0.9784 & 0.1710 \\ -0.9811 & 0.0858 & -0.1736 \end{bmatrix}\begin{bmatrix} 0.948 \\ 5.924 \\ 0 \end{bmatrix} = \begin{bmatrix} -0.966 \\ -5.904 \\ -0.422 \end{bmatrix}$$

$$\Rightarrow \quad \bar{V} = -0.966\,\bar{I} - 5.904\,\bar{J} - 0.422\,\bar{K}\,\frac{\text{km}}{\text{s}}$$

<u>Two-Line Element Sets</u>

The North American Aerospace Defense Command (NORAD) developed the Two-Line Element (TLE) format for transmitting satellite orbital elements of thousands of space objects daily.

TLEs are provided in a structured format intended to be very compact. The format is difficult to read unless you are familiar with the structure.

TLEs provide twelve parameters for each space object as follows:

- <u>Six</u> classical orbital elements

    <u>Four</u> Keplerian elements $e$, $i$, $\Omega$, $\omega$

    <u>Two</u> additional elements

    - $n$, the mean motion (mean value), substituted for $a$
    - $M$, the mean anomaly, substituted for $\nu$

- <u>Three</u> parameters, which describe the effects of perturbations on satellite motion

    - $B^*$, $\frac{1}{2}\left(\frac{dn}{dt}\right)$, $\frac{1}{6}\left(\frac{d^2n}{dt^2}\right)$

- <u>Two</u> parameters for identification purposes

- <u>One</u> parameter giving the epoch time the data is recorded

A sample TLE and an explanation is provided by Ref. [2] below

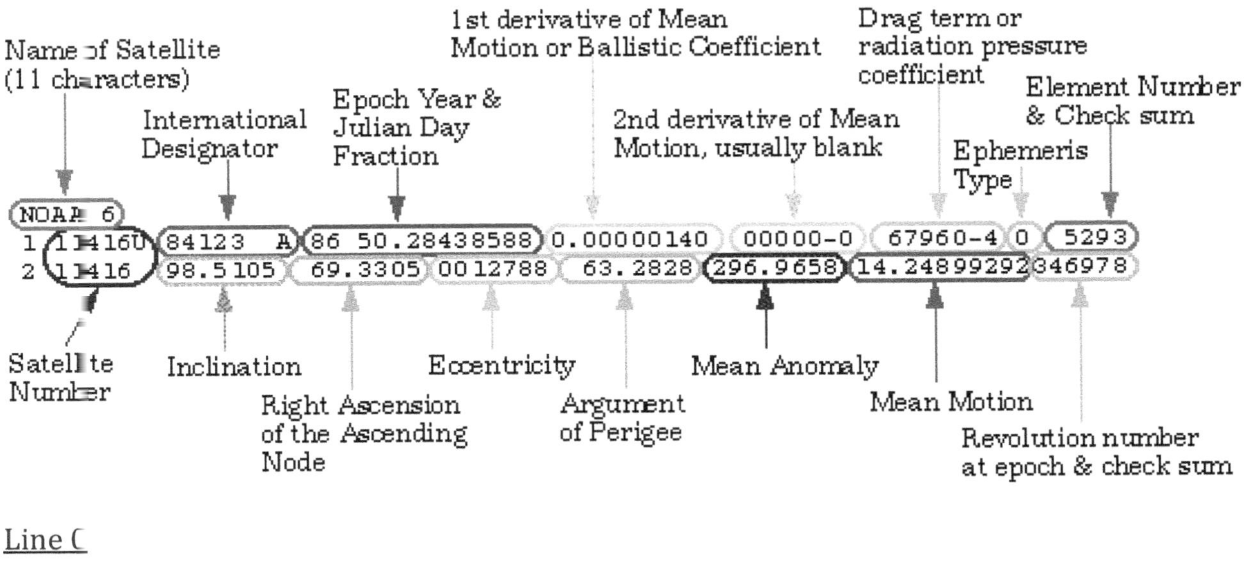


<u>Line 0</u>

Name of Satellite:

(NOAA 6): This is simply the name associated with the satellite.

Line 1

International Designator:

(84123 A): The 84 indicates launch year was in 1984, while the 123 means it's the 124th launch of the year, and "A" shows it was the first object resulting from this launch.

Epoch Date and Julian Date Fraction:

The Julian day fraction is the number of days passed in the particular year. For example, the date above shows "86" as the epoch year (1986) and 50.28438588 as the Julian day fraction meaning a little over 50 days after January 1, 1986. The resulting time of the vector would be 1986, 050:06:49:30.94. This was computed as

- Start with 50.28438588 days (Days = 50)
- 50.28438588 days − 50 = 0.28438588 days
- 0.28438588 days × 24 hrs/day = 6.8253 hours (Hours = 6)
    - 6.8253 hours − 6 = 0.8253 hours
    - 0.8253 hours × 60 min/hr = 49.5157 minutes (Minutes = 49)
    - 49.5157 − 49 = 0.5157 minutes
    - 0.5157 minutes × 60 sec/min = 30.94 seconds (Seconds = 30.94)

Ballistic Coefficient:

(0.00000140): This is also called the first derivative of mean motion. The ballistic coefficient is the daily rate of change in the number of revs the object completes each day, divided by 2. Units are revs/day. This is "catch all" drag term used in the Simplified General Perturbations (SGP4) predictor of the United States Space Command (USSPACECOM).

Second Derivative of Mean Motion:

(00000-0 = 0.00000): The second derivative of mean motion is a second order drag term in the SGP4 predictor used to model terminal orbit decay. It measures the second time derivative in daily mean motion, divided by 6. Units are $(\text{revs/day})^3$. A leading decimal must be applied to this value. The last two characters define an applicable power of 10. (12345-5 = 0.0000012345).

Drag Term:

(67960-4 = 0.000067960): This is also called the radiation pressure coefficient (or BSTAR). The parameter is another drag term in the SGP4 predictor. Units are $(\text{earth radii})^{-1}$. The last two characters define an applicable power of 10. Do not confuse this parameter with "B-Term", the USSPACECOM special perturbations factor of drag coefficient, multiplied by reference area, divided by weight.

Element Set Number and Check Sum:

(5293): The element set number is a running count of all two-line element sets generated by USSPACECOM for this object (in this example, 529). Since multiple agencies perform this function, numbers are skipped on occasion to avoid ambiguities. The counter should always increase with time until it exceeds 999, when it reverts to 1. The last number of the line is the check sum of line 1.

<u>Line 2</u>

Satellite Number:

(11416): This is the catalog number USSPACECOM has designated for this object. A "U" indicates an unclassified object.

Inclination (degrees):

This is the angle between the equatorial and the orbital planes. The value provided is the True Equator, Mean Equinox (TEME) mean inclination.

Right Ascension of the Ascending Node (degrees):

The angle between the vernal equinox and the point where the orbit crosses the equatorial plane (moving north). The value provided is the TEME right ascension of the ascending node.

Eccentricity:

(0012788): This is a constant defining the shape of the orbit (0 = circular, less than 1 = elliptical). The value provided is the mean eccentricity. A leading decimal must be applied to this value.

Argument of Perigee (degrees):

This is the angle between the ascending node and the orbit's point of closest approach to Earth (perigee). The value provided is the TEME mean argument of perigee.

Mean Anomaly (degrees):

This is the angle, measured from perigee, of the satellite location in the orbit referenced to a circular orbit with radius equal to the semi-major axis.

Mean Motion:

(14.24899292): This is the value is the mean number of orbits per day the object completes. There are 8 digits after the decimal, leaving no trailing space(s) when the following element exceeds 9999.

Revolution Number and Check Sum:

(346978): This is the orbit number at Epoch Time, chosen very near the time of true ascending node passage as a matter of routine. The last digit is the check sum for line 2.

Problems

11.1 Find the ECI position and velocity vectors for the Keplerian orbital elements below:

$a = 10{,}000.0$ km  $e = 0.080$  $i = 40.0^{\circ}$  $\Omega = 50.0^{\circ}$  $\omega = 60.0^{\circ}$  $\nu = 70.0^{\circ}$

(Ans. $\bar{R} = -8{,}343.59\,\bar{I} - 1{,}114.15\,\bar{J} + 4{,}762.22\,\bar{K}$ km,
$\bar{V} = -1.16\,\bar{I} - 5.93\,\bar{J} - 2.45\,\bar{K}\,\frac{km}{s}$)

11.2 Given the orbital elements

$a = 9{,}000.0$ km $\quad e = 0.050 \quad i = 40.0^o \quad \Omega = 45.0^o \quad \omega = 50.0^o \quad \nu = 20.0^o$

Determine the corresponding ECI position and velocity vectors.

(Ans. . $\bar{R} = -2{,}290.83\,\bar{I} + 6{,}438.28\,\bar{J} + 5{,}179.27\,\bar{K}$ km,
$\bar{V} = -5.96\,\bar{I} - 3.26\,\bar{J} + 1.60\,\bar{K}\ \frac{km}{s}$)

11.3 Calculate the ECI position and velocity vectors for the orbital elements below:

$a = 9{,}400.0$ km $\quad e = 0.015 \quad i = 164.0^o \quad \Omega = 199.0^o \quad \omega = 310.0^o \quad \nu = 206.0^o$

(Ans. . $\bar{R} = 7{,}015.98\,\bar{I} + 6{,}355.01\,\bar{J} + 1{,}068.01\,\bar{K}$ km,
$\bar{V} = 4.28\,\bar{I} - 4.51\,\bar{J} - 1.62\,\bar{K}\ \frac{km}{s}$)

11.4 Consider the low-Earth satellite having the following orbital elements:

$a = 6{,}652.556$ km $\quad e = 0.04 \quad i = 30.0^o \quad \Omega = 10.0^o \quad \omega = 20.0^o \quad \nu = 30.0^o$

Compute the ECI position and velocity vectors for the satellite.

(Ans. . $\bar{R} = 3{,}324.17\,\bar{I} + 4{,}910.65\,\bar{J} + 2{,}458.82\,\bar{K}$ km,
$\bar{V} = -6.74\,\bar{I} + 3.45\,\bar{J} + 2.64\,\bar{K}\ \frac{km}{s}$)

11.5 For the orbital elements of an Earth satellite given below, find the ECI position and velocity vectors.

$a = 10{,}550$ km $\quad e = 0.06 \quad i = 102^o \quad \Omega = 327^o \quad \omega = 272^o \quad \nu = 345^o$

(Ans. $\bar{R} = -778.3\,\bar{I} + 2{,}905.5\,\bar{J} - 9{,}469.9\,\bar{K}$ km, $\bar{V} = 5.50\,\bar{I} - 3.23\,\bar{J} - 1.34\,\bar{K}\ \frac{km}{s}$)

11.6 For the orbital elements of an Earth satellite given below, calculate the ECI position and velocity vectors.

$a = 9{,}081.88$ km $\quad e = 0.2227 \quad i = 32.45^o \quad \Omega = 107.57^o \quad \omega = 72.44^o$
$\nu = 299.738^o$

(Ans. $\bar{R} = -1{,}342.48\,\bar{I} + 8{,}236.33\,\bar{J} - 767.06\,\bar{K}$ km, $\bar{V} = -5.67\,\bar{I} - 2.09\,\bar{J} + 3.84\,\bar{K}\ \frac{km}{s}$)

11.7 A tracking station is observing a foreign satellite in an elliptical orbit and determined the vehicle has the following ECI position and velocity vectors given below.

$$\bar{R} = 13{,}863.18\,\bar{I} - 1{,}183.06\,\bar{J} - 435.81\,\bar{K}\text{ km} \qquad \bar{V} = -1.99\,\bar{I} - 4.53\,\bar{J} - 0.59\,\bar{K}\,\frac{\text{km}}{\text{s}}$$

Determine the following quantities
(a) Keplerian orbital elements of the satellite at its initial position
(b) time since this satellite passed perigee
(c) orbital elements of the satellite at a position 34 min and 53.2 sec after it leaves its initial position
(d) ECI position and velocity vectors of the satellite at its second position.
(Ans. (a) $a = 12{,}300$ km, $e = 0.34$, $i = 172.0^\circ, \Omega = 188.0^\circ, \omega = 323.0^\circ, \nu = 230.0^\circ$;
(b) $t - t_0 = 166.676$ min; (c) $\nu = 79.52^\circ$);
(d) $\bar{R} = -8{,}451.4\,\bar{I} + 5{,}739.0\,\bar{J} + 963.6\,\bar{K}$ km, $\bar{V} = 1.98\,\bar{I} + 6.38\,\bar{J} + 0.85\,\bar{K}\,\frac{\text{km}}{\text{s}}$)

11.8 A radar sighting of an Earth satellite determines that it has the following ECI position and velocity vectors at time 1.

$$\bar{R} = 5{,}247.08\,\bar{I} - 10{,}122.31\,\bar{J} + 3{,}442.35\,\bar{K}\text{ km}$$
$$\bar{V} = 4.90\,\bar{I} - 1.94\,\bar{J} + 2.13\,\bar{K}\,\frac{\text{km}}{\text{s}}$$

Compute the following
(a) the Keplerian orbital elements of the satellite
(b) time since this satellite passed the perigee of the orbit
(c) the orbital elements of the satellite at position 2, 45 minutes after it leaves position 1
(d) the ECI position and velocity vectors of the satellite at position 2.
(Ans. (a) $a = 11{,}500$ km, $e = 0.04$, $i = 152.0^\circ, \Omega = 208.0^\circ, \omega = 296.0^\circ, \nu = 206.0^\circ$;
(b) $t - t_0 = 118.22$ min; (c) $\nu = 282.8^\circ$,
(d) $\bar{R} = 10{,}786.1\,\bar{I} - 1{,}399.6\,\bar{J} - 3{,}349.5\,\bar{K}, \bar{V} = -1.59\,\bar{I} - 5.33\,\bar{J} - 2.11\,\bar{K}\,\frac{\text{km}}{\text{s}}$)

# Module 12: Launching

When a space vehicle is launched, many factors affect the resulting orbit. These factors include the latitude of the launch site, the launch direction, or launch azimuth, the velocity needed for orbit insertion, and the rotation of Earth. The effect that these factors have on the resulting orbit are addressed below.

## Launch Azimuth

The projection of the path of a vehicle launched from a launch site located at a latitude, $\phi$ onto the surface of Earth forms the spherical triangle in the figure below.

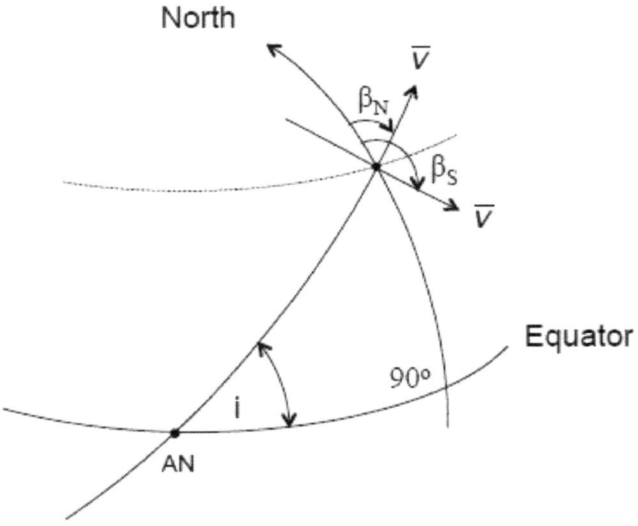


The orbital inclination resulting from the launch is shown as angle, i, and the inertial velocity of the vehicle is shown as, $\overline{V}$. The angle, $\beta$, is known as the heading, or launch azimuth, which is measured clockwise from north. Using spherical trigonometry, the relationship between launch azimuth, inclination, and the launch site latitude is

$$\cos i = \cos\phi \sin\beta$$

Knowing the launch site latitude and the launch azimuth, $\beta$ ,the resulting orbital inclination can be determined by the expression

$$i = \cos^{-1}[\cos\phi \sin\beta]$$

The orbital inclination is minimized when the term sin $\beta$ is maximized, which occurs when $\beta = 90^o$, or the launch is due east, giving the relationship

$$i_{MIN} = \cos^{-1}[\cos\phi]$$

Therefore, the minimum inclination possible is equal to the latitude of the launch site. As a result, an equatorial orbit can only be achieved from a launch site located on the equator.

Based upon the launch site latitude, the resulting orbital inclinations are

| | |
|---|---|
| $\underline{\beta = 90^o}$: | $i_{MIN} = \phi$ (for direct orbits) |
| | $i_{MIN} = 180^o - \phi$ (retrograde orbits) |
| For all other $\beta$: | $i > i_{MIN}$ |
| For a given $\phi$: | $\phi \leq i \leq 180^o - \phi$ |

Knowing the launch site latitude and the desired orbital inclination, the required launch azimuth can be determined using the formula

$$\beta = \sin^{-1}\left[\frac{\cos i}{\cos\phi}\right]$$

This gives two possible solutions, one northern launch opportunity, and one southern launch opportunity, as

$$\beta_N = \sin^{-1}\left[\frac{\cos i}{\cos\phi}\right]$$

$$\beta_S = 180^o - \beta_N$$

These two launch opportunities are also shown on the figure above. Both launch directions will result in the same orbital inclination, the only difference being a shift in the longitude of the ascending node of the two orbits. The southern launch gives a smaller value of $\Omega$ than a northern launch, i.e., the AN for a southern launch will be west of the AN for a northern launch.

The result of a southern launch is that the orbital plane will be rotated clockwise about the Z-axis from the orbital plane obtained from a southern launch.

Launches in the northeast or southeast direction result in a direct orbit, while launches in the northwest or southwest direction result in a retrograde orbit.

One widely-used launch site in the United States is John F. Kennedy Space Center (KSC) located on Merritt Island, Florida has a latitude of approximately 28.5° N and a longitude of 80.65° W. Due to geographical considerations launches from KSC are restricted to launch azimuths given by

$$35^{\circ} \leq \beta \leq 120^{\circ}$$

KSC is ideal for launching in a due east direction resulting in an orbital inclination of 28.5°.

A second widely-used launch site in the U.S. is Vandenberg Space Force Base (VBG), near Lompoc in Santa Barbara County, CA, which has a latitude of approximately 34.63° N and a longitude of 120° W. Launches from VBG are restricted to launch azimuths given by

$$140^{\circ} \leq \beta \leq 201^{\circ}$$

VBG is ideal for southerly launches resulting in a polar orbit with an inclination of 90°.

<u>Example 12.1</u>

Find the possible launch azimuths from a launch site located at a latitude of 15° N resulting in the following orbital inclinations
(a) 57°
(b) 142°
(c) 10°.

Solution:

$$\text{(a)}\ \beta_N = \sin^{-1}\left[\frac{\cos i}{\cos\phi}\right] = \sin^{-1}\left[\frac{\cos 57^{\circ}}{\cos 15^{\circ}}\right] = 34.3^{\circ}\quad\text{(northeast)}$$

$$\beta_S = 180^{\circ} - \beta_N = 180^{\circ} - 34.3^{\circ} = 145.6^{\circ}\quad\text{(southeast)}$$

$$\text{(b)}\ \beta_N = \sin^{-1}\left[\frac{\cos i}{\cos\phi}\right] = \sin^{-1}\left[\frac{\cos 142^{\circ}}{\cos 15^{\circ}}\right] = -54.7^{\circ} + 360^{\circ} = 305.3^{\circ}\quad\text{(northwest)}$$

$$\beta_S = 180^{\circ} - \beta_N = 180^{\circ} - 305.3^{\circ} = -125.3^{\circ} + 360^{\circ} = 234.7^{\circ}\quad\text{(southeast)}$$

$$\text{(c)}\ \beta_N = \sin^{-1}\left[\frac{\cos i}{\cos\phi}\right] = \sin^{-1}\left[\frac{\cos 10^{\circ}}{\cos 15^{\circ}}\right] = \infty\quad\text{(not possible)}$$

$$\beta_S = 180^{\circ} - \beta_N = 180^{\circ} - \infty = -\infty\quad\text{(not possible)}$$

Example 12.2

Determine the possible and permissible launch azimuths from KSC which will result in the following orbital inclinations
(a) 38°
(b) 55°
(c) 17.5°.

Solution:

At KSC, $\phi = 28.5^o$, and the permissible launch azimuths are $35^o \leq \beta \leq 120^o$

(a) $i = 38^o$:

$$\beta_N = \sin^{-1}\left[\frac{\cos i}{\cos \phi}\right] = \sin^{-1}\left[\frac{\cos 38^o}{\cos 28.5^o}\right] = 63.7^o \;\Rightarrow \text{possible and permissible}$$

$$\beta_S = 180^o - \beta_N = 180^o - 63.7^o = 116.3^o \quad \Rightarrow \text{possible and permissible}$$

(b) $i = 55^o$:

$$\beta_N = \sin^{-1}\left[\frac{\cos i}{\cos \phi}\right] = \sin^{-1}\left[\frac{\cos 55^o}{\cos 28.5^o}\right] = 40.7^o \;\Rightarrow \text{possible and permissible}$$

$$\beta_S = 180^o - \beta_N = 180^o - 40.7^o = 139.3^o \quad \Rightarrow \text{possible, \underline{not} permissible}$$

(c) $i = 17.5^o$:

$$\beta_N = \sin^{-1}\left[\frac{\cos i}{\cos \phi}\right] = \sin^{-1}\left[\frac{\cos 17.5^o}{\cos 28.5^o}\right] = \infty \quad \Rightarrow \text{\underline{not} possible}$$

$$\beta_S = 180^o - \beta_N = 180^o - \infty = -\infty \quad \Rightarrow \text{\underline{not} possible}$$

Launch Site Dynamics

A launch site at a random location in the Northern Hemisphere is shown on the figure below. The position vector of the launch site, $\bar{r}_{LS}$, and the velocity of the launch site, $\bar{v}_{LS}$, are indicated. Both vectors are expressed in the ECI coordinate system.

The velocity of the launch site is shown to be in the easterly direction due to the CCW rotation of Earth.

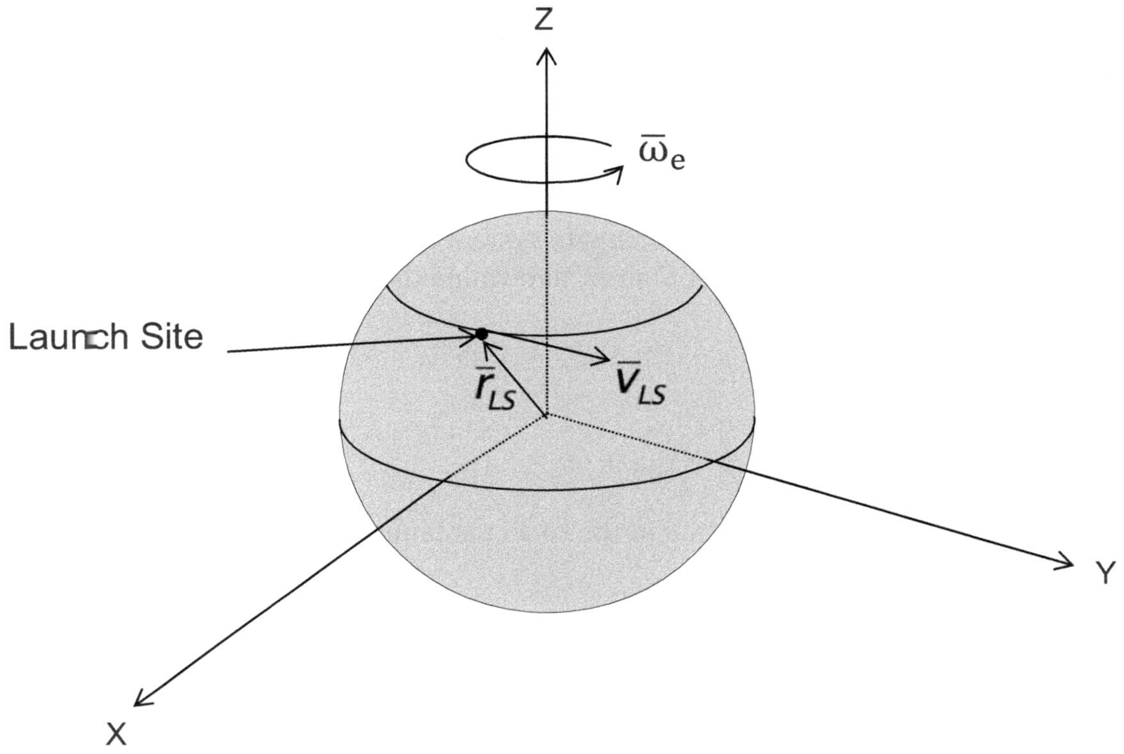


Because of Earth rotation, the launch site and vehicle have an initial velocity prior to launch. The velocity of the launch site is

$$\bar{V}_{LS} = \bar{\omega}_e \times \bar{r}_{LS}$$

This velocity must be included in computing the vehicle's inertial velocity, $\bar{V}$, as

$$\bar{V} = \bar{V}_{LS} + \bar{v}_{rel}$$

$$\Rightarrow \quad \bar{V} = \bar{\omega}_e \times \bar{r}_{LS} + \bar{v}_{rel}$$

where

$\bar{v}_{rel} =$ velocity of vehicle relative to launch site immediately after launch

The acceleration of the launch site will be negligible compared to the thrust acceleration of the launch vehicle.

Example 12.3

A rocket is launched from launching site located on the equator and 30° east of the ECI X-axis. Consider it has an initial velocity and acceleration relative to the launch site as

$$\bar{v}_{rel} = 0.5\,\bar{\imath} + 1.0\,\bar{\jmath}\,\frac{km}{s} \qquad \bar{a}_{rel} = 1.1\,\bar{\imath} + 0.9\,\bar{\jmath}\,\frac{km}{s^2}$$

Assuming that the origin of the relative coordinate system is centered at the launch site and the Cartesian axes are parallel with the ECI axes, determine the inertial velocity and acceleration of the rocket at launch.

Solution:

$$\bar{V} = \dot{\bar{\rho}} + \bar{v}_{rel} + \bar{\omega}_e \times \bar{r} = \bar{\omega}_e \times \bar{\rho} + \bar{v}_{rel} + \bar{\omega}_e \times \bar{r}$$

Assuming that the relative position of the rocket at launch is $\bar{r} = 0$.

$$\bar{V} = \dot{\bar{\rho}} + \bar{v}_{rel} = \bar{\omega}_e \times \bar{\rho} + \bar{v}_{rel}$$

$$\bar{V} = \bar{\omega}_e \times (\rho \cos 30^\circ\,\bar{I} + \rho \sin 30^\circ\,\bar{J}) + 0.5\,\bar{I} + 1.0\,\bar{J}$$

$$\bar{V} = (7.292 \times 10^{-5}\,\bar{K}) \times [(6{,}378.1) \cos 30^\circ\,\bar{I} + (6{,}378.1) \sin 30^\circ\,\bar{J}] + 0.5\,\bar{I} + 1.0\,\bar{J}$$

$$\bar{V} = 0.40\,\bar{J} - 0.23\,\bar{I} + 0.5\,\bar{I} + 1.0\,\bar{J}$$

$$\Rightarrow \quad \bar{V} = 0.27\,\bar{I} + 1.40\,\bar{J}\,\frac{km}{s}$$

$$\ddot{\bar{A}} = \ddot{\bar{\rho}} + \bar{a}_{rel} + 2 \times \bar{v}_{rel} + \bar{\alpha} \times \bar{r} + \bar{\omega}_e \times (\bar{\omega}_e \times \bar{r})$$

$$\ddot{\bar{A}} = \bar{\omega}_e \times (\bar{\omega}_e \times \bar{\rho}) + \bar{a}_{rel} + 2(\bar{\omega}_e \times \bar{v}_{rel})$$

$$\ddot{\bar{A}} = (7.292 \times 10^{-5}\,\bar{K}) \times [(7.292 \times 10^{-5}\,\bar{K}) \times (6{,}378.1) \cos 30^\circ\,\bar{I} + (6{,}378.1) \sin 30^\circ\,\bar{J}]$$
$$+ (1.1\,\bar{I} + 0.9\,\bar{J}) + 2[(7.292 \times 10^{-5}\,\bar{K}) \times (0.5\,\bar{I} + 1.0\,\bar{J})]$$

$$\Rightarrow \quad \bar{A} \approx 1.1\,\bar{I} + 0.9\,\bar{J}\,\frac{km}{s^2} = \bar{a}_{rel}$$

## Ground Tracks

The projection of a satellite's orbit onto Earth's surface is called its ground track and is generally displayed on a Mercator projection, as shown by the series of figures below which were provided by Ref. [3].

The orbital plane of an Earth satellite passes through the center of Earth as shown below.

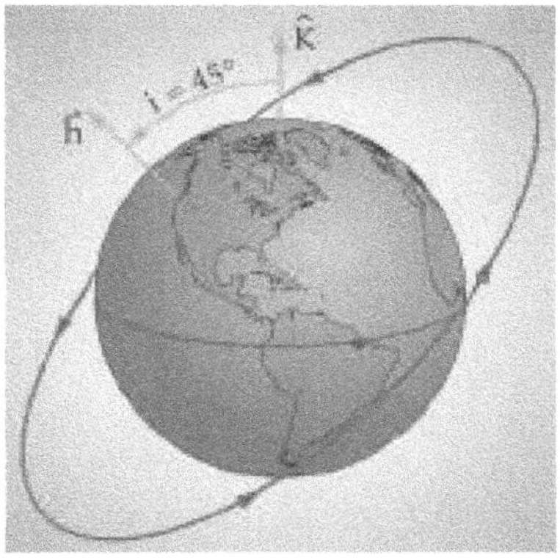


If Earth didn't rotate, the ground track would be a great circle on Earth's surface and would continue to retrace itself.

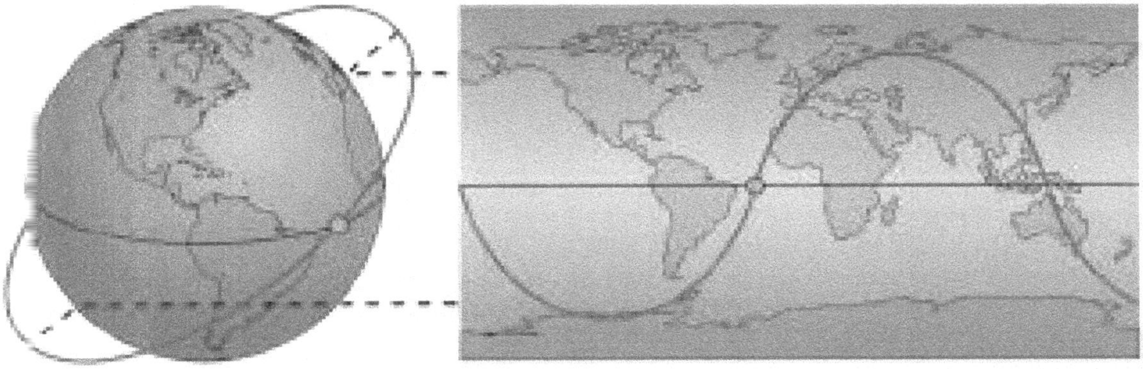

Because the satellite reaches a maximum and minimum latitude while passing over the equator twice, the ground track of a low-Earth satellite will resemble a sine wave, as shown above.

However, because Earth rotates eastward beneath the satellite's orbit at the rate of 15.04° per hour, the ground track moves westward at that same rate.

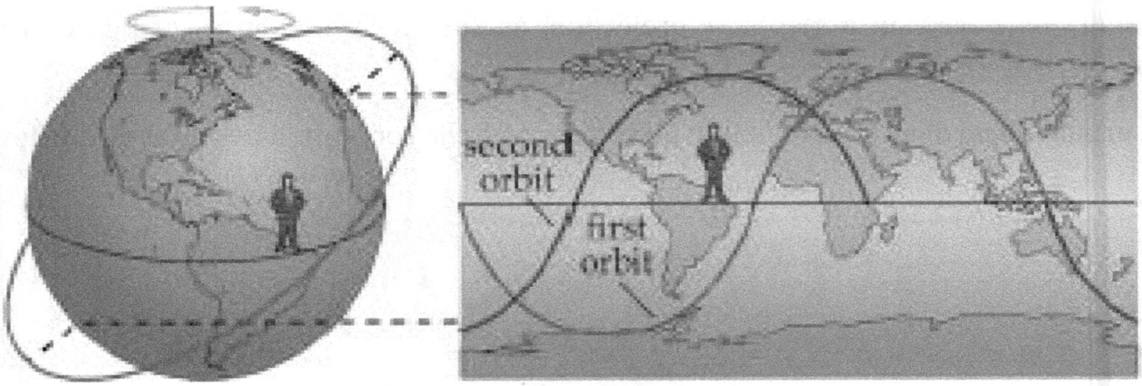


The appearance of ground tracks are affected by

- orbital period
- inclination
- eccentricity
- altitude
- direction of travel
- perigee/apogee locations

Orbital period:

- The period can be determined from the distance (in degrees longitude) between successive ground tracks, as

  $\Rightarrow$ T = distance/15.04° per hour (in hrs.).

- After determining the period, the semi-major axis of the orbit can be found using the expression

$$a = \sqrt[3]{\frac{\mu T^2}{4\pi^2}}$$

- Orbital periods, which are exact multiples of a sidereal day (23h, 56 m, 4.09 s), create ground tracks which repeat themselves at regular intervals, e.g., 1-day repeating, 3-day repeating, 4-day repeating, etc.

  ⟹ good for flying over the same areas on a regular basis.

- Non-repeating orbits will produce a solid band of coverage between the maximum north and south latitudes over time.

Inclination:

- The maximum north and south latitudes which a ground track reaches indicates the inclination of the satellite's orbit.

- As the value of inclination increases, the ground tracks will become steeper and appear more 'squared' at the peaks and valleys of the projection as shown below.

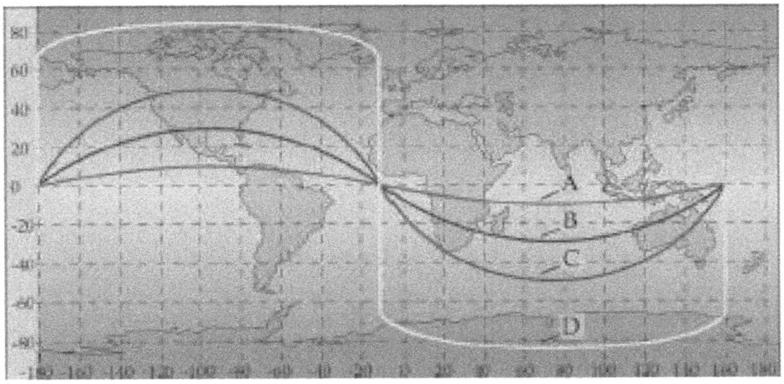


- In the figure above, all four ground tracks represent orbits with a period of 4 hours. The inclination of these orbits is determined by looking at the highest latitude reached. Orbit A has an inclination of 10°. Orbit B has an inclination of 30°. Orbit C has an inclination of 50°. Orbit D has an inclination of 85°. (Note that Orbit D appears distorted because ground distances elongate near the poles on a Mercator projection.)

Eccentricity:

- If a ground track is symmetrical about a longitude and resembles a sine wave,

$$\Rightarrow \quad e = 0 \quad \text{(circular orbit)}$$

- Otherwise, if the ground track is unsymmetrical the orbit is elliptical.

$$\Rightarrow \quad 0 < e < 1 \text{ (elliptical orbit)}$$

- By varying the eccentricity, ground tracks can be tilted relative to a longitude.

Altitude:

- As the altitude increases, the orbital period becomes larger, the more compact the ground track becomes greatly affecting the appearance of ground tracks having the same inclination, as shown below.

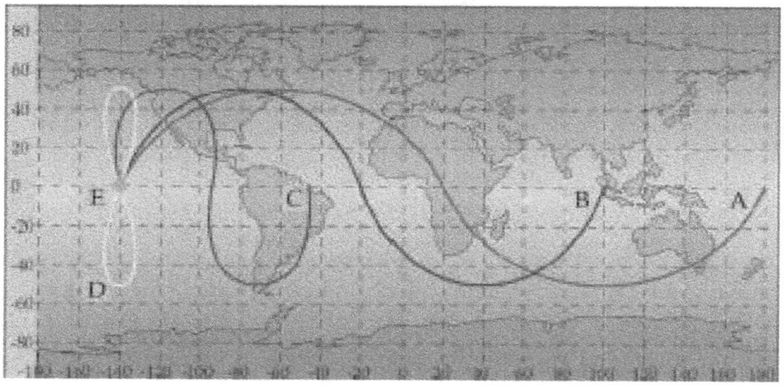


- In this figure, Orbit A has a period of 2.67 hours. Orbit B has a period of 8 hours. Orbit C has a period of 18 hours. Orbit D has a period of 24 hours. Orbit E has a period of 24 hours.

- Geosynchronous orbit is an orbit having a period equal to the rotation rate of Earth = 23 h 56 m 4.09 s (one sidereal day)

$\Rightarrow$ ground track will continuously repeat itself

If $i \neq 0^\circ \Rightarrow$ ground track will look like a figure 8, spending an equal amount of time in the Northern and Southern Hemispheres.

If $i = 0^\circ \Rightarrow$ geostationary orbit, ground track will be a stationary point on the equator.

Direction of travel:

- If <u>any</u> part of a ground track proceeds eastward, the orbit is direct.
- Otherwise, the orbit could be either direct or retrograde.
- Very high altitude, direct orbits exhibit sharpened peaks and valleys.

For LEO, if one orbital period spans

$< 360^{\circ}$ of Earth longitude $\Longrightarrow$ direct orbit.

$> 360^{\circ}$ of Earth longitude $\Longrightarrow$ retrograde orbit.

Perigee/apogee location:

- By altering the perigee and apogee locations in an orbit, the ground track can be made more unsymmetrical, as shown below.

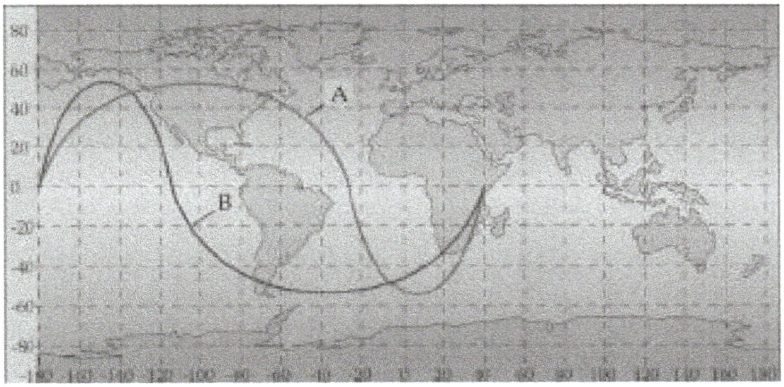


- Both ground tracks have orbits with periods of 11.3 hours, inclinations of 50°, and high eccentricities. Orbit A has perigee located over the Northern Hemisphere, while B has perigee over the Southern Hemisphere.

Orbits can be designed to create many different types of ground tracks to meet specific mission objectives.

Problems

12.1 Find the possible and permissible launch azimuths for a launch from KSC into orbits having the following inclinations
(a) 29°
(b) 54°
(c) 15.5°.
(Ans. (a) $\beta_N = 84.4°, \beta_S = 95.6°$; (b) $\beta_N = 42.0°$; (c) none)

12.2 Calculate the possible and permissible launch azimuths for a launch from VBG having the following inclinations
(a) 58°
(b) 152.5°
(c) 100°
(Ans. (a) none; (b) none; (c) $\beta_S = 192.2°$)

12.3 Possible launch azimuths for a launch from VBG for a particular mission range from 150°to 190°. Determine the possible range of orbital inclinations for this range of launch azimuths.
(Ans. $65.7° \leq i \leq 98.2°$)

12.4 It is desired to launch a satellite directly into an orbit having an inclination of 116.57°. Calculate the possible launch azimuths at launch sites located at each of the following latitudes
(a) $\phi = 28.5°$
(b) $\phi = 34.5°$
(c) $\phi = 5.5°$.
(Ans. (a) $\beta_N = 329.4°, \beta_S = 210.6°$; (b) $\beta_N = 327.1°, \beta_S = 212.9°$; (c) $\beta_N = 333.3°$ , $\beta_S = 206.7°$)

12.5 A tourist at KSC notices a satellite passing directly overhead and determines the approximate heading to be 341°.
(a) What is the approximate inclination of the satellite?
(b) If the tracking station at KSC can track objects in low-Earth orbit which come within 4° latitude of the tracking station, for what range of orbital inclinations will this tracking station be useful?
(Ans. (a) $i = 106.6°$; (b) $24.5° \leq i \leq 155.5°$)

12.6 A launch site is located on the Greenwich Meridian at a latitude of 30° N and the Greenwich Meridian is aligned with the vernal equinox direction. At launch, a rocket has an inertial velocity vector given as

$$\bar{V} = 0.2\,\bar{I} + 1.0\,\bar{J} + 0.1\,\bar{K}\ \frac{km}{s}$$

Calculate the velocity vector of the rocket relative to the launch site at launch.
(Ans. $\bar{v}_{rel} = 0.2\,\bar{I} + 0.6\,\bar{J} + 0.1\,\bar{K}\ \frac{km}{s}$)

12.7 Consider a rocket launched from VBG having an initial velocity of 3.0 $\frac{km}{s}$ in the northerly direction relative to the launch site. Determine the inertial velocity of the rocket at the time of launch. Assume the Greenwich Meridian is aligned with the vernal equinox direction at the launch time.
(Ans. $\bar{V} = 1.19\,\bar{I} + 1.28\,\bar{J} + 2.47\,\bar{K}\ \frac{km}{s}$)

12.8 For the launch scenario described in Problem 12.7, find the inertial acceleration of the rocket if its relative acceleration at launch is 2.0 $\frac{km}{s^2}$.
(Ans. $\bar{A} = 0.58\,\bar{I} + 0.98\,\bar{J} + 1.65\,\bar{K}\frac{km}{s^2}$)

(This page was intentionally left blank.)

# Module 13: Single-Impulse Maneuvers - Changing the Size and Shape of an Orbit

Orbital maneuvers are very important in orbital mechanics. In this discussion of orbital maneuvers, all velocity changes are considered to be impulsive, i.e., the velocity change, $\Delta\bar{v}$, applied to a vehicle will occur instantaneously at a point with no change in position.

The assumption that velocity changes are impulsive, provides very accurate results for preliminary analysis and mission planning activities. For highly accurate analyses, the equations of motion must be integrated numerically during the time the $\Delta\bar{v}$ takes place, which could be seconds, minutes, or hours.

A tangential $\Delta\bar{v}$ in the direction of travel applied at one end of an elliptical orbit will change the altitude at the other end resulting in a change to both a and e to change. Therefore, if the $\Delta\bar{v}$ is applied at periapsis, the altitude at apoapsis will be increased, and vice-versa.

Since velocity changes are considered impulsive, the altitude will not change at the point of $\Delta\bar{v}$ application, and due to constant energy, the point of application will remain on the new orbit.

Consider the elliptical orbit 1 with the periapsis velocity $\bar{v}_p$

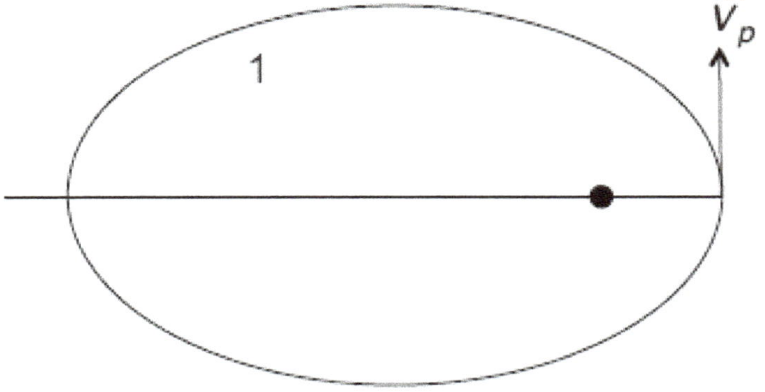

A tangential $\Delta\bar{v}$ in the direction of travel at the periapsis of orbit 1 raises the altitude at the apoapsis

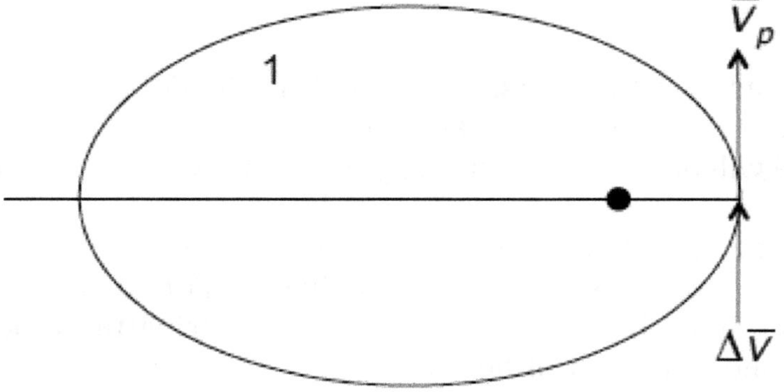


This maneuver creates a new orbit 2 having velocity $\bar{v}_p'$ at the periapsis of orbit 2

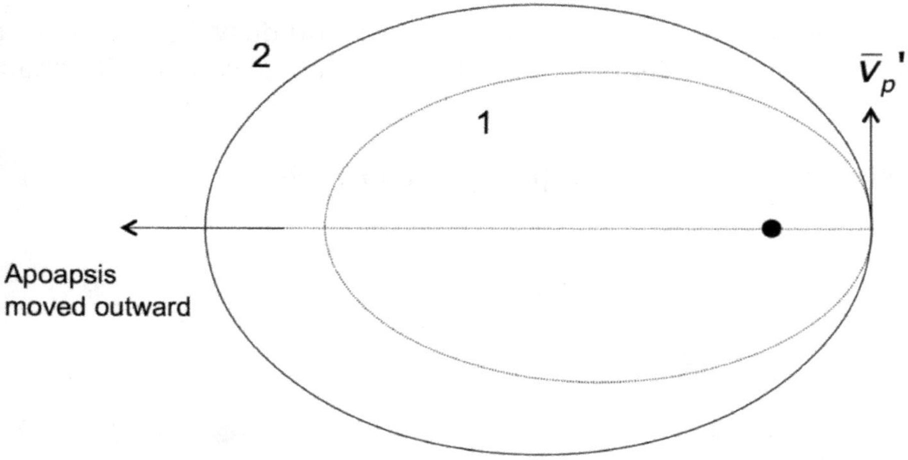

Similarly, a tangential $\Delta\bar{v}$ in the opposite direction of travel at the periapsis of orbit 1 lowers the altitude at the apoapsis of orbit 1

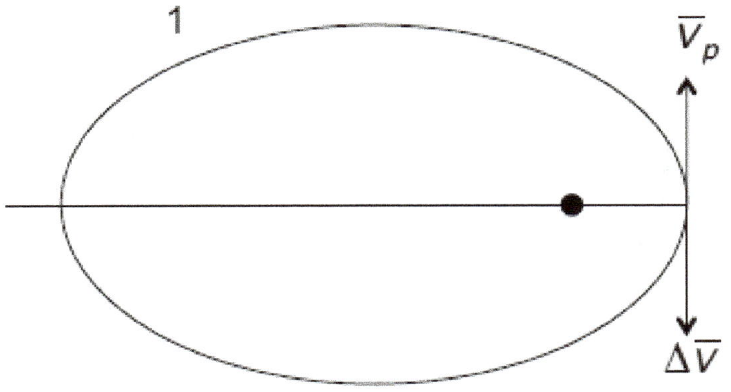


This creates a new orbit 2 with a new velocity $\bar{v}_p'$ at the periapsis of orbit 2

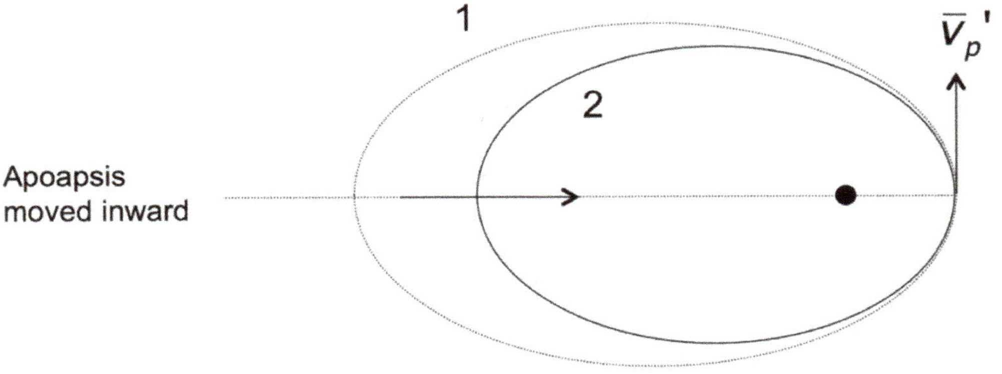


Consider the elliptical orbit 1 with the apoapsis velocity $\bar{v}_a$

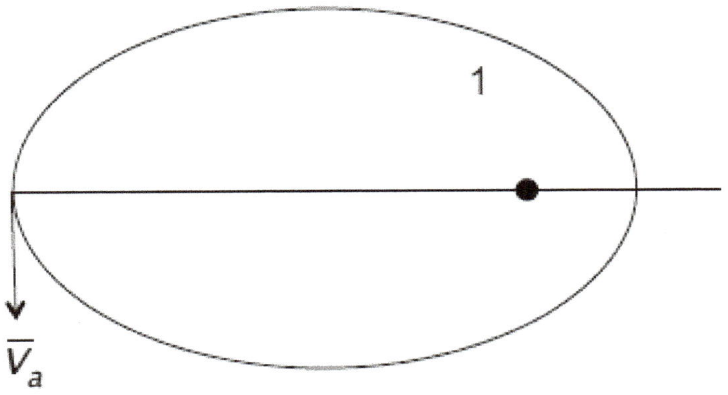

A tangential $\Delta\bar{v}$ in the direction of travel at apoapsis raises the altitude at the periapsis of orbit 1

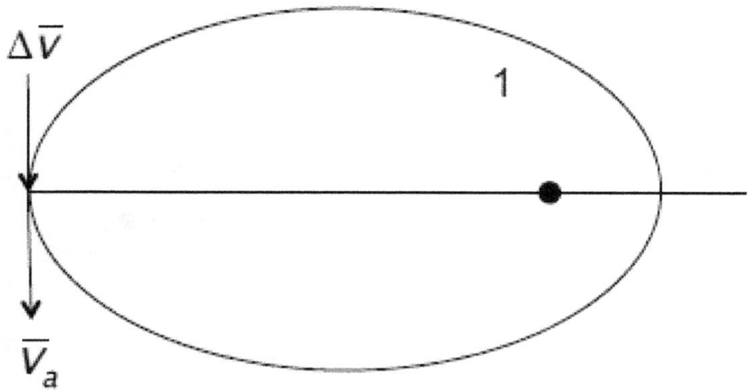


This creates the new elliptical orbit 2 having a velocity at apoapsis of $\bar{v}_a'$

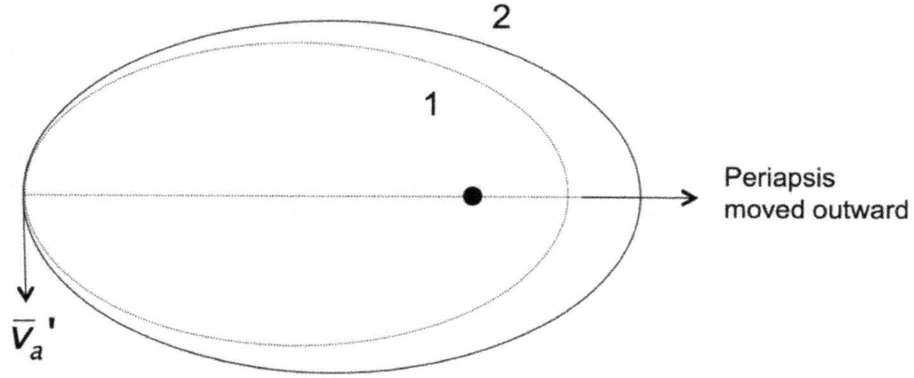


Consider the elliptical orbit 1 with the apoapsis velocity $\bar{v}_a$

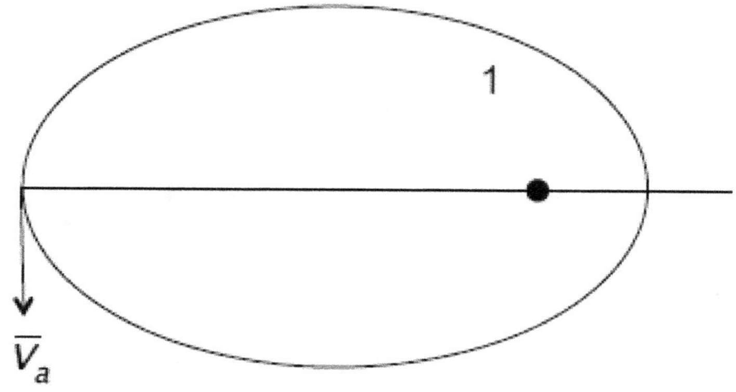

A tangential $\Delta\bar{v}$ applied in the opposite direction of travel at the apoapsis of orbit 1 will lower the altitude at the periapsis of orbit 1

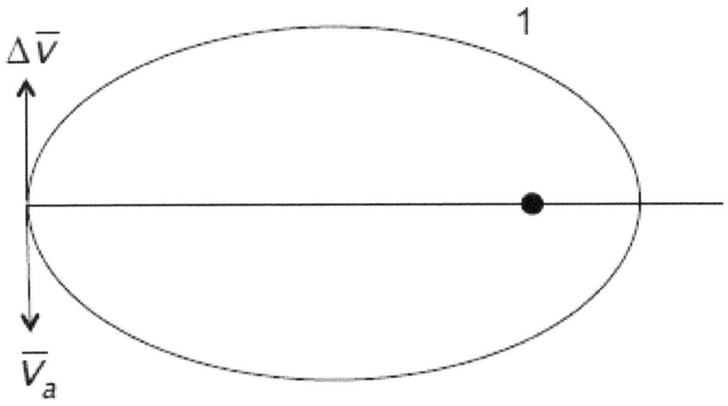


This creates a new orbit 2 having the velocity at the apoapsis, $\Delta\bar{v}_a'$

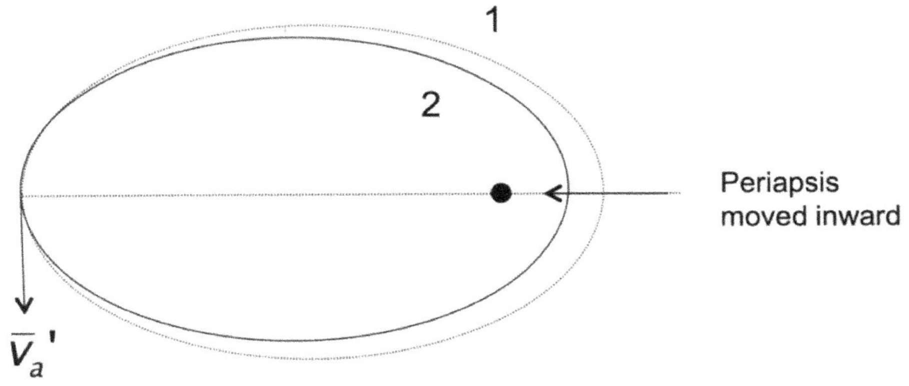


<u>Example 13.1</u>

An Earth satellite is in a circular orbit at an altitude of 350 km. If an onboard rocket can provide a velocity increase of 0.5 $\frac{km}{s}$, calculate the semi-major axis and the speed at the apogee of the satellite's new orbit.

Solution:

$$r_p = R_E + (alt)_p = 6{,}378.1 + 350 = 6{,}728.1 \text{ km}$$

$$v = \sqrt{\frac{\mu}{r}} = \sqrt{\frac{3.986\times10^5}{6{,}728.1}} = 7.70 \ \frac{km}{s}$$

$$v_p = v + \Delta v = 7.697 + 0.5 = 8.20\ \frac{km}{s}$$

$$h = r_p v_p = (6{,}728.1)(8.20) = 55{,}170.4\ \frac{km^2}{s}$$

$$r_p = \frac{\frac{h^2}{\mu}}{1+e}$$

$$e = \frac{h^2}{\mu r_p} - 1 = \frac{(55{,}170.4)^2}{(3.986\times10^5)(6{,}728.1)} - 1 = 0.135$$

$$r_a = \frac{\frac{h^2}{\mu}}{1-e} = \frac{(55{,}170.4)^2}{(3.986\times10^5)(1-0.135)} = 8{,}827.9$$

$$a = \frac{r_a}{1+e} = \frac{8{,}827.9}{1+0.135}$$

$$\Rightarrow\ a = 7{,}777.9\ km$$

$$v_a = \sqrt{\mu\left(\frac{2}{r_a} - \frac{1}{a}\right)} = \sqrt{(3.986\times10^5)\left(\frac{2}{8{,}827.9} - \frac{1}{7{,}777.9}\right)}$$

$$\Rightarrow\ v_a = 6.25\ \frac{km}{s}$$

Example 13.2

A spacecraft is in a circular orbit at an altitude of 300 km. An onboard thruster provides an instantaneous velocity change that results in a maximum altitude of 900 km during the spacecraft's next orbit. Determine the velocity change required to perform this maneuver.

Solution:

$$r_p = R_E + (alt)_p = 6{,}378.1 + 300 = 6.678.1\ km$$

$$v_c = \sqrt{\frac{\mu}{r_p}} = \sqrt{\frac{3.986\times10^5}{6{,}678.1}} = 7.73\ \frac{km}{s}$$

$$r_a = R_E + (alt)_a = 6{,}378.1 + 900 = 7{,}278.1\ km$$

$$a = \frac{r_a + r_p}{2} = \frac{7{,}278.1 + 6{,}678.1}{2} = 6{,}978.1\ km$$

$$v_p = \sqrt{\mu\left(\frac{2}{r_p} - \frac{1}{a}\right)} = \sqrt{(3.986 \times 10^5)\left(\frac{2}{6{,}678.1} - \frac{1}{6{,}978.1}\right)} = 7.89\ \frac{km}{s}$$

$$\Delta v = v_p - v_c = 7.89 - 7.73\ \frac{km}{s}$$

$$\Rightarrow\ \Delta v = 0.16\ \frac{km}{s}$$

Another common application of a single-impulse maneuver is circularizing an orbit. This process is described below.

Consider an elliptical orbit 1 with an apoapsis velocity $\bar{v}_a$

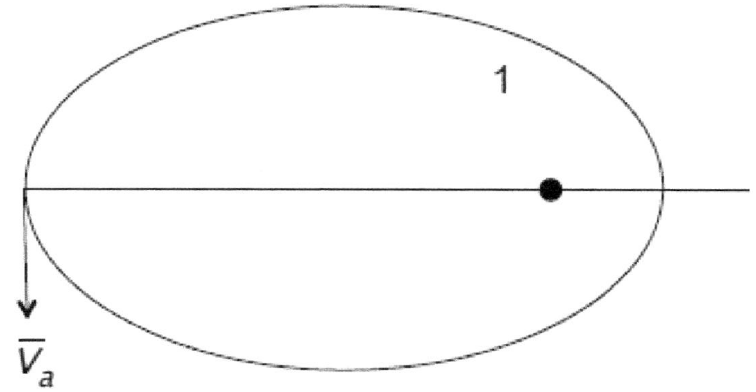


A tangential $\Delta\bar{v}$ can be applied at the apoapsis of orbit 1 in the direction of travel to raise the periapsis altitude to a value equal to the apoapsis altitude

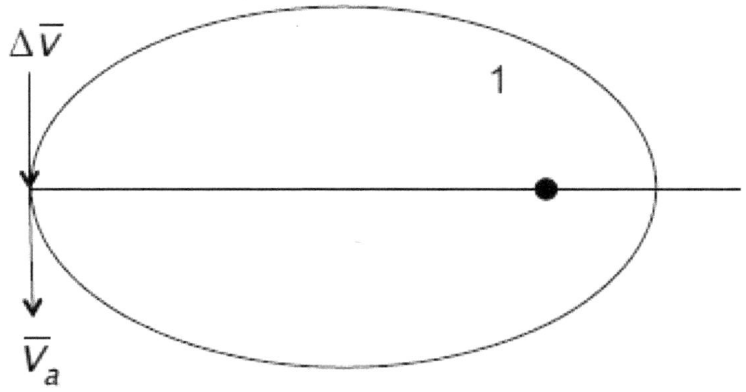

This creates a circular orbit 2 having a circular velocity $\bar{v}_c$

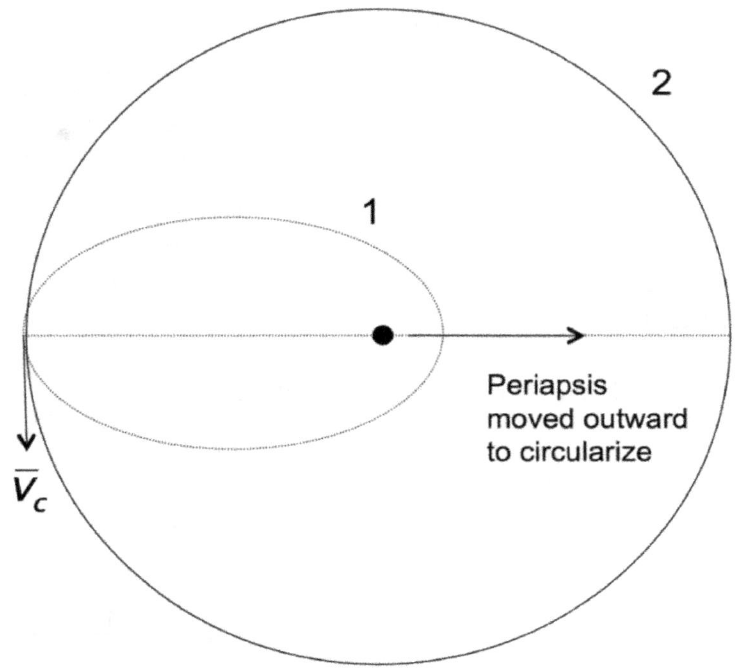


To circularize an orbit at periapsis altitude, consider an elliptical orbit 1 with a periapsis velocity $\bar{v}_p$

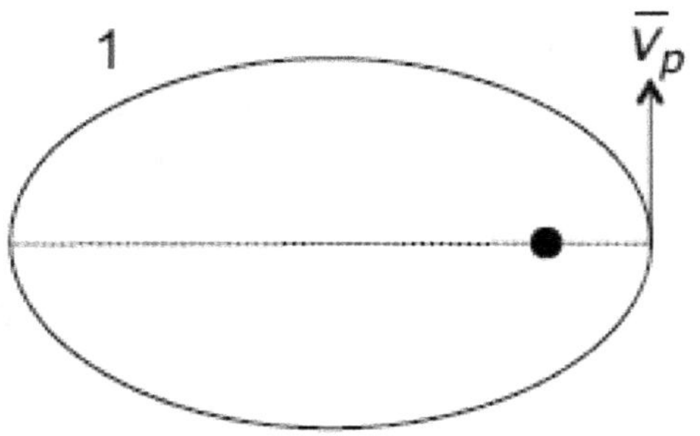

A tangential $\Delta\bar{v}$ van be applied at the periapsis of orbit 1 in the opposite direction of travel to lower the apoapsis altitude to a value equal to the periapsis altitude

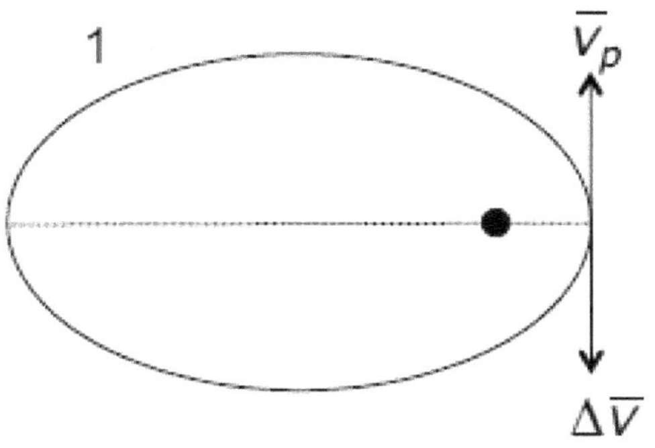


This creates a circular orbit 2 having a circular velocity $\bar{v}_c$

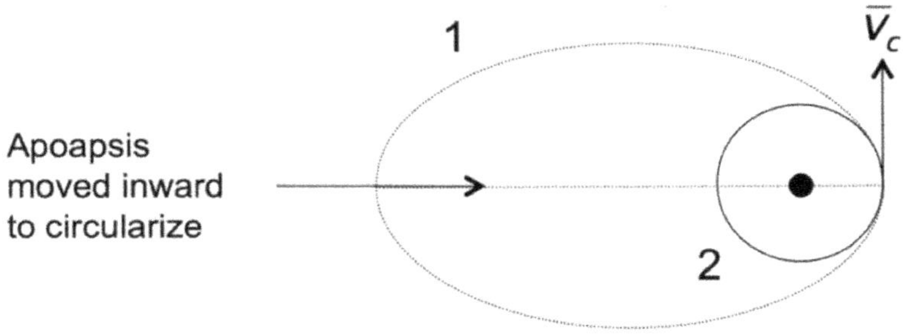


<u>Example 13.3</u>

A satellite has a speed of 8.1 $\frac{km}{s}$ at an altitude of 500 km at perigee. Calculate the $\Delta v$, to be applied at apogee, to circularize its orbit at the apogee altitude.

Solution:

$$r_p = R_E + (alt) = 6{,}378.1 + 500 = 6{,}878.1 \text{ km}$$

$$E = \frac{v^2}{2} - \frac{\mu}{r_p} = \frac{(8.1)^2}{2} - \frac{3.986\times10^5}{6{,}878.1} = -25.15\ \frac{km^2}{s^2}$$

$$a = \frac{-\mu}{2E} = \frac{-3.986\times10^5}{2(-25.15)} = 7{,}924.5 \text{ km}$$

$$a = \frac{r_p + r_a}{2}$$

$$r_a = 2a - r_p = 2(7{,}924.5) - 6{,}878.1 = 8{,}970.8\ \text{km}$$

$$v_a = \sqrt{\mu\left(\frac{2}{r_a} - \frac{1}{a}\right)} = \sqrt{(3.986 \times 10^5)\left(\frac{2}{8{,}970.8} - \frac{1}{7{,}924.5}\right)} = 6.21\ \frac{\text{km}}{\text{s}}$$

$$v_c = \sqrt{\frac{\mu}{r_a}} = \sqrt{\frac{3.986\times 10^5}{8{,}970.8}} = 6.67\ \frac{\text{km}}{\text{s}}$$

$$\Delta v = v_c - v_a = 6.67 - 6.21$$

$$\Rightarrow\ \Delta v = 0.46\ \frac{\text{km}}{\text{s}}$$

Problems

13.1 A spacecraft is in a 700 km circular Earth orbit. A $\Delta v$ of $0.75\ \frac{\text{km}}{\text{s}}$ is applied in the direction of travel. Find the semi-major axis and eccentricity of the new orbit. (Ans. $a = 8{,}957.2$ km, $e = 0.21$)

13.2 A satellite is in a circular Earth orbit at an altitude of 200 km. As a certain point a thruster is fired and the speed of the satellite is increased instantaneously by 40%. If this point becomes the perigee of the new orbit, calculate the speed of the satellite in the new orbit when the true anomaly has a value of $120^\circ$.
(Ans. $v = 5.46\ \frac{\text{km}}{\text{s}}$)

13.3 An Earth satellite is orbiting such that $a = 12{,}000$ km and $e = 0.2$. Determine the $\Delta v$ necessary such that the altitude at periapsis is increased by 400 km.
(Ans. $\Delta v = 0.06\ \frac{\text{km}}{\text{s}}$)

13.4 Consider an Earth orbiting spacecraft having $a = 9{,}000$ km, $e = 0.1$. Find the velocity required to circularize the orbit at the periapsis altitude.
(Ans. $\Delta v = -\ 0.33\ \frac{\text{km}}{\text{s}}$)

13.5 For an Earth orbiting spacecraft having $a = 11{,}000$ km, $e = 0.3$. Calculate the velocity required to circularize the orbit at the apoapsis altitude.
(Ans. $\Delta v = 0.86\ \frac{\text{km}}{\text{s}}$)

13.6 For the case of a satellite having $a = 7{,}000$ km and $e = 0$, if the orbital speed is increased by 40%, determine the size and shape of the resulting orbit.
(Ans. $a = 678{,}132.2$ km, $e = 0.99$)

13.7 The shape of a circular orbit can be changed to one having an eccentricity of 0.1 by either increasing or decreasing the satellite's speed. Calculate the velocity change required (in terms of the circular orbital speed) for both cases.
(Ans. $\Delta v_1 = 0.05\ v_c\ , \Delta v_2 = -\ 0.05\ v_c$)

13.8 A space vehicle orbits Earth with a semi-major axis of 7,000 km and eccentricity of 0.0317. Find the minimum velocity change required to change the eccentricity to 1.0.
(Ans. $\Delta v = 3.06\ \frac{km}{s}$)

(This page was intentionally left blank.)

# Module 14: Single-Impulse Maneuvers – Planetary Escape, Capture, and Re-Entry

Planetary escape, planetary capture, and planetary re-entry can all be accomplished using a single-impulse. These three maneuvers are described in the paragraphs below.

Consider circular orbit 1 with a circular velocity $\bar{v}_c$

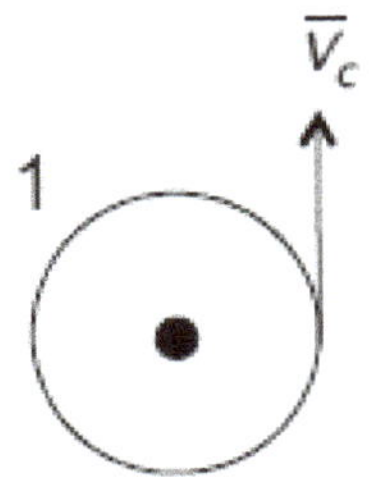


A tangential $\Delta\bar{v}$ can be calculated to increase the velocity to equal the escape velocity

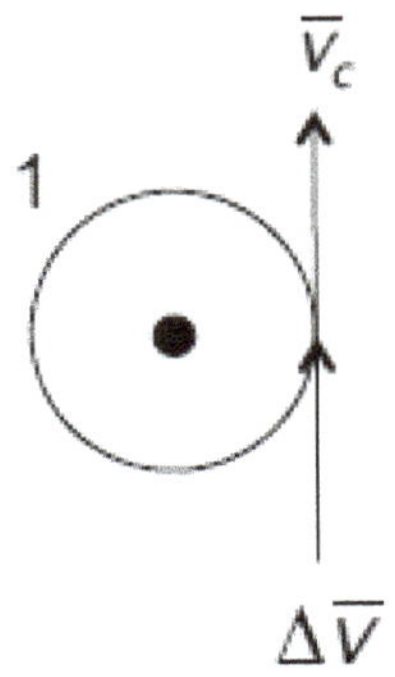

This creates the escape trajectory 2 with the escape velocity, $\overline{v}_{esc}$ at the point of application of $\Delta\overline{v}$ as

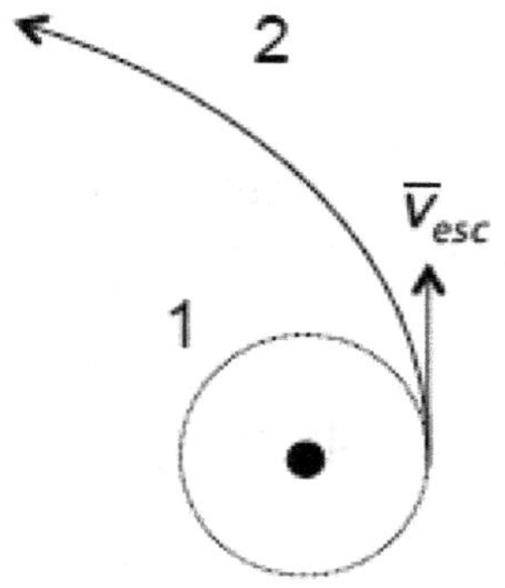


Consider a parabolic approach trajectory 1 with a velocity at the point of closest approach of $\overline{v}$

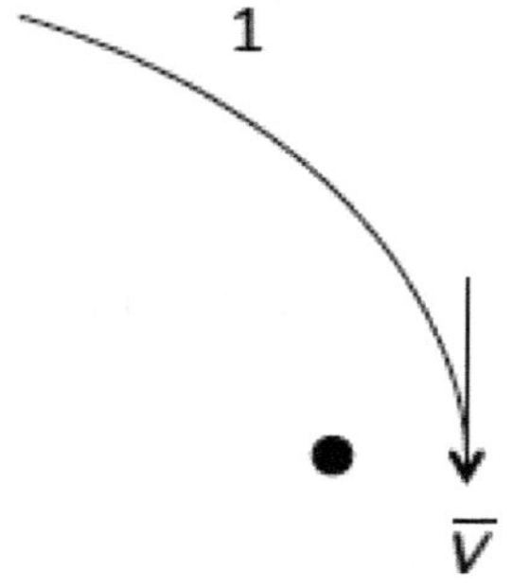


A tangential $\Delta\overline{v}$ can be applied in the direction opposite to the velocity to reduce the speed and allow capture by the planet in a circular orbit

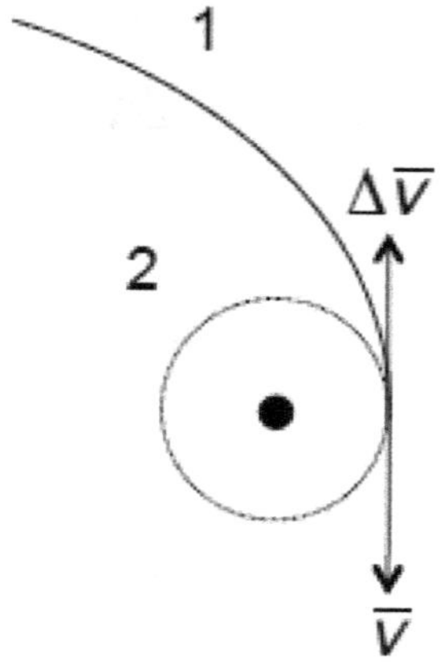

Therefore, capture is achieved in orbit 2 with a circular velocity $\bar{v}_c$

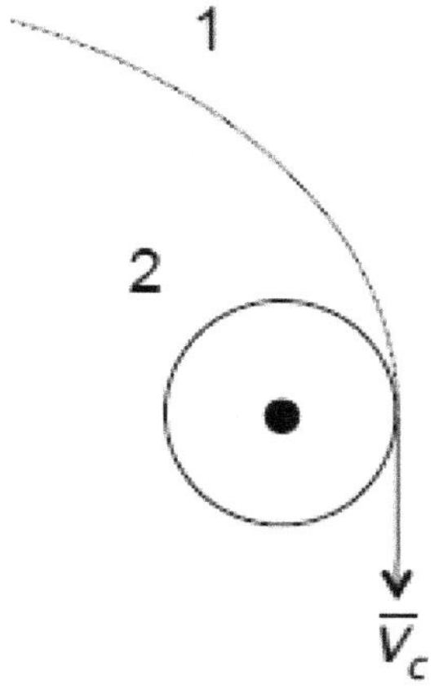


Consider the circular orbit 1 having velocity $\bar{v}_c$

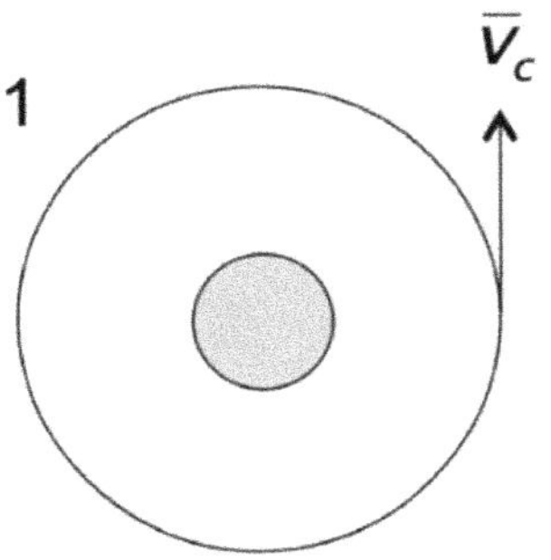


For planetary re-entry the velocity must be reduced to re-enter the atmosphere and affect landing or such that the new trajectory intersects the surface of the planet. To do so, a tangential $\Delta\bar{v}$ is applied in the opposite direction of the velocity

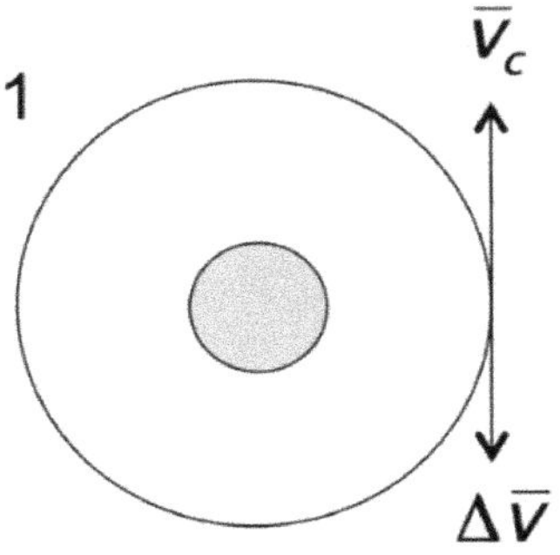

Once the velocity is reduced to $\bar{v}_{Re}$, the spacecraft will move from circular orbit 1 to the elliptical re-entry trajectory 2, as

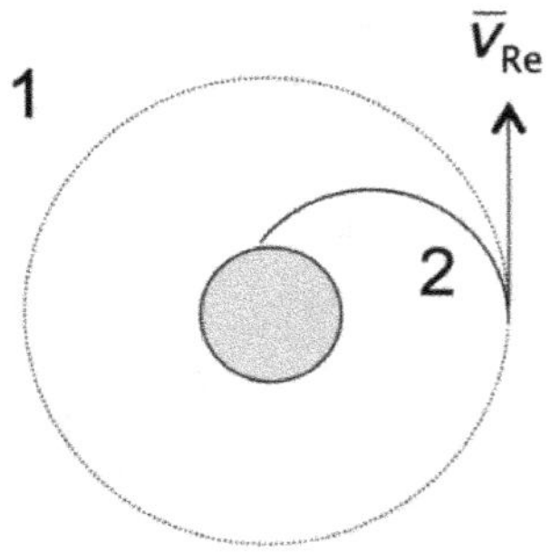


If no further maneuvers are undertaken, and neglecting the effects of any atmosphere, the spacecraft will impact the surface of the planet with velocity $\bar{v}$

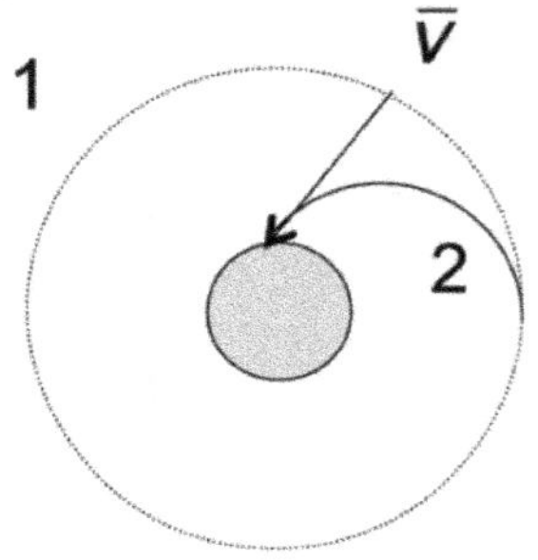


Example 14.1

A space probe is in a circular orbit of a distant planet at r = 5,800 km and is designed to make a hard landing on the planet's surface. A single $\Delta v$ of $0.65 \frac{km}{s}$ is applied to reduce the probe's speed and cause de-orbit. The point of application of the $\Delta v$ becomes the apoapsis of the re-entry ellipse such that surface impact will occur at the point where the true anomaly is 244.9°. Assuming that the planet has no atmosphere, calculate the radius of the planet. Use: $\mu_{PL} = 2.67 \times 10^5 \frac{km}{s}$.

Solution:

$$v_c = \sqrt{\frac{\mu_{PL}}{r}} = \sqrt{\frac{2.67\times10^5}{5,800}} = 6.78 \frac{km}{s}$$

$$v = v_c - \Delta v = 6.78 - 0.65 = 6.13 \frac{km}{s}$$

$$a = \frac{1}{\frac{2}{r} - \frac{v^2}{\mu_{PL}}} = \frac{1}{\frac{2}{5,800} - \frac{(6.13)^2}{2.67 \times 10^5}} = 4,899.7 \text{ km}$$

$$e = \frac{r}{a} - 1 = \frac{5,800}{4,899.7} - 1 = 0.184$$

$$R = \frac{a(1-e^2)}{1+e\cos\nu} = \frac{(4,899.7)\left[1-(0.184)^2\right]}{1+(0.184)\cos 244.9^o}$$

$$\Rightarrow \quad R = 5,134.6 \text{ km}$$

<u>Example 14.2</u>

In Example 14.1, if R = 5,000 km and $\Delta v = 0.55 \frac{km}{s}$ find the true anomaly at impact.

Solution:

$$v_c = \sqrt{\frac{\mu_{PL}}{r}} = \sqrt{\frac{2.67 \times 10^5}{5,800}} = 6.78 \ \frac{km}{s}$$

$$v = v_c - \Delta v = 6.78 - 0.55 = 6.23 \ \frac{km}{s}$$

$$a = \frac{1}{\frac{2}{r} - \frac{v^2}{\mu_{PL}}} = \frac{1}{\frac{2}{5,800} - \frac{(6.23)^2}{2.67 \times 10^5}} = 5,013.5 \text{ km}$$

$$e = \frac{r}{a} - 1 = \frac{5,800}{5,013.5} - 1 = 0.157$$

$$R = \frac{a(1-e^2)}{1+e\cos\nu}$$

$$\nu = \cos^{-1}\left\{\frac{1}{e}\left[\frac{a(1-e^2)}{R} - 1\right]\right\} = \cos^{-1}\left\{\frac{1}{(0.157)}\left[\frac{5,013.5(1-(0.157)^2)}{5,000.0} - 1\right]\right\}$$

$$\Rightarrow \quad \nu = 261.9^o$$

Example 14.3

A spacecraft in a circular orbit (1) of radius r leaves for infinity on a parabolic trajectory (2) and returns from infinity on another parabolic trajectory (3) to a circular orbit (4) having radius 12r. Find the total Δv required (in terms of μ and r for this transfer.

Solution:

Orbit 1, circular velocity at r: $v_1 = \sqrt{\frac{\mu}{r}}$

Orbit 2, escape velocity at departure: $v_2 = \sqrt{\frac{2\mu}{r}}$

Orbit 3, escape velocity at return : $v_3 = \sqrt{\frac{2\mu}{12r}}$

Orbit 4, circular velocity at 12r: $v_4 = \sqrt{\frac{\mu}{12r}}$

$$\Delta v_1 = v_2 - v_1 = \sqrt{\frac{2\mu}{r}} - \sqrt{\frac{\mu}{r}} = (\sqrt{2} - 1)\sqrt{\frac{\mu}{r}} = 0.414\sqrt{\frac{\mu}{r}}$$

$$\Delta v_2 = v_4 - v_3 = \sqrt{\frac{2\mu}{12r}} - \sqrt{\frac{\mu}{12r}} = \frac{(\sqrt{2}-1)}{\sqrt{12}}\sqrt{\frac{\mu}{r}} = 0.120\sqrt{\frac{\mu}{r}}$$

$$\Delta v = \Delta v_1 + \Delta v_2 = 0.414\sqrt{\frac{\mu}{r}} + 0.120\sqrt{\frac{\mu}{r}}$$

$$\Rightarrow \Delta v = 0.534\sqrt{\frac{\mu}{r}}$$

Problems

14.1 A spacecraft in a 500 km circular Earth orbit is to use a single $\Delta v$ to de-orbit and land at a point where the true anomaly of the de-orbit ellipse has a value of $300^{\circ}$. Calculate the velocity change that will accomplish this mission.
(Ans. $\Delta v = -0.19\ \frac{km}{s}$)

14.2 A satellite traveling to the moon on a parabolic trajectory is to be captured into a circular lunar orbit at its point of closest approach, 400 km. Find the velocity change necessary at the moon to accomplish the maneuver.
(Ans. $\Delta v = 0.63\ \frac{km}{s}$)

14.3 A starship is in a circular parking orbit about planet M having

$$\mu = 4.0 \times 10^{6}\ \frac{km^3}{s^2} \qquad a = 12{,}000\ km \qquad e = 0.3$$

If the starship is to leave this orbit from periapsis on a parabolic escape trajectory, determine the increase in velocity required for departure.
(Ans. $\Delta v = 5.98\ \frac{km}{s}$)

14.4 Calculate the $\Delta v$ necessary for a space probe approaching Mars on a parabolic trajectory to be captured at the periapsis of an elliptical orbit having a semi-major axis of 6,000 km and an eccentricity of 0.2.
(Ans. $\Delta v = 0.95\ \frac{km}{s}$)

14.5 In Problem 14.4, find the $\Delta v$ required for the probe to be captured at the apoapsis of the Mars orbit rather than the periapsis.
(Ans. $\Delta v = 1.27\ \frac{km}{s}$)

14.6 A spacecraft in a circular Earth has an altitude of 600 km. Compute the velocity change needed for it to depart on a hyperbolic trajectory if the required hyperbolic excess speed is $3.2\ \frac{km}{s}$.
(Ans. $\Delta v = 3.59\ \frac{km}{s}$)

14.7 A space probe is in a circular orbit ($r = 5{,}800$ km) about a planet where $\mu = 2.67 \times 10^5 \frac{km^3}{s^2}$. A single $\Delta v$ of $0.6 \frac{km}{s}$ is applied to reduce the vehicle's speed and de-orbit the probe and the point of application of the velocity change becomes the apoapsis of the re-entry ellipse. Determine the true anomaly of the impact point if the radius of the planet is 5,400 km.
(Ans. $\nu = 236.1°$)

14.8 An Earth satellite is placed in a circular polar orbit at an altitude of 400 km. As the satellite passes over the North Pole a retro-rocket is fired which produces a burst of negative thrust that instantaneously reduces its speed to a value which will ensure an equatorial landing. Calculate the required speed change to accomplish this task.
(Ans. $\Delta v = -0.23 \frac{km}{s}$)

# Module 15: Single-Impulse Maneuvers - Rotating the Line-of-Apsides

One of the more creative single-impulse maneuvers is a rotation of the line-of-apsides, i. e., the apse line, of an orbit. The most common rotations are either 180° or 90°, which are described below, but other rotations are also possible.

## Rotating the Line-of-Apsides by 180°

Consider the rotation of the line-of-apsides, or flipping the periapsis and apoapsis, in an elliptical orbit by 180°, i.e., moving from orbit 1 to orbit 2, as shown

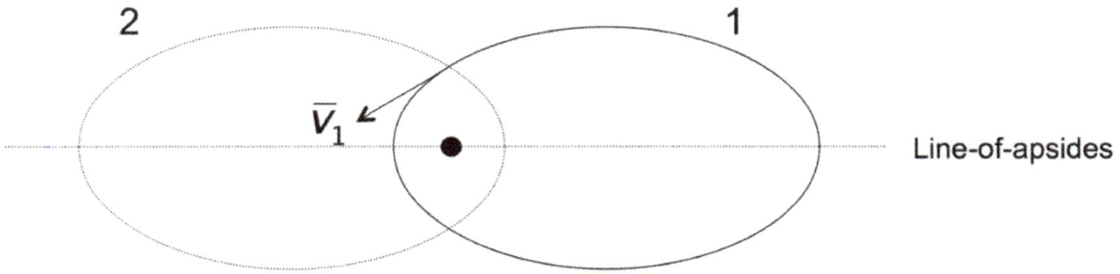


The velocities at the intersection of orbit 1 and orbit 2 are equal in magnitude but have different directions. A $\Delta\bar{v}$ must be applied at the intersection to rotate $\bar{v}_1$ to orbit 2 and transfer to the new elliptical orbit

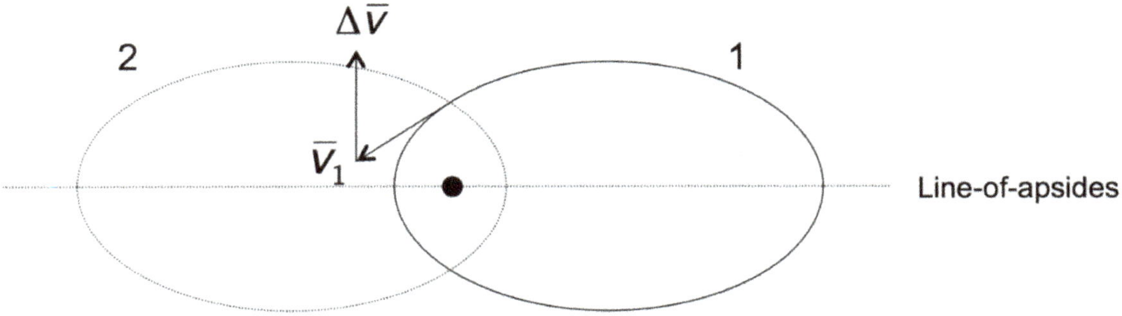

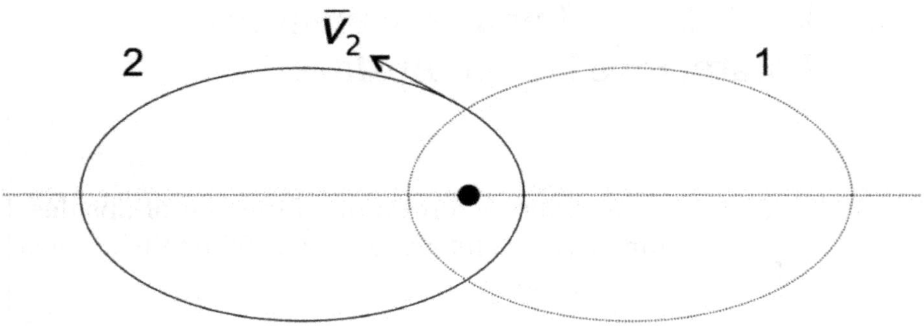


Example 15.1

For the 180° rotation of the apse line shown above, calculate the Δv required in the case where a = 8,000 km and e = 0.2 for an Earth orbit.

Solution:

$$r = \frac{a(1-e^2)}{1+e\cos\nu} = \frac{8{,}000\,[1-(0.2)^2]}{1+(0.2)\cos 90^\circ} = 7{,}680 \text{ km} = p$$

$$h = \sqrt{\mu p} = \sqrt{(3.986 \times 10^5)(7{,}680)} = 55{,}328.5 \ \frac{\text{km}^2}{\text{s}}$$

$$v = \sqrt{\mu\left(\frac{2}{r} - \frac{1}{a}\right)} = \sqrt{(3.986 \times 10^5)\left(\frac{2}{7{,}680} - \frac{1}{8{,}000}\right)} = 7.35 \ \frac{\text{km}}{\text{s}}$$

$$\varphi = \cos^{-1}\left(\frac{h}{rv}\right) = \cos^{-1}\left[\frac{55{,}328.5}{7{,}680(7.35)}\right] = 11.43^\circ$$

$$\Delta v = 2v\sin\varphi = 2(7.35)\sin 11.43^\circ$$

$$\Rightarrow \ \Delta v = 2.91 \ \frac{\text{km}}{\text{s}}$$

Example 15.2

In Example 15.1, use the Law of Cosines to solve for the Δv required for the transfer.

Solution:

$$v_1 = v_2 = 7.35\ \frac{km}{s}$$

$$\varphi = 11.43^o$$

$$\Delta v = \sqrt{v_1^2 + v_2^2 - 2v_1 v_2 \cos 2\varphi}$$

$$\Delta v = \sqrt{(7.35)^2 + (7.35)^2 - 2(7.35)(7.35)\cos[(2)(11.43^o]}$$

$$\Rightarrow \Delta v = 2.91\ \frac{km}{s}$$

Rotating the Line-of-Apsides by 90°

Consider the rotation of the line-of-apsides, or rotating the apoapsis in an elliptical orbit, by 90°, i.e., moving from orbit 1 to orbit 2, as shown

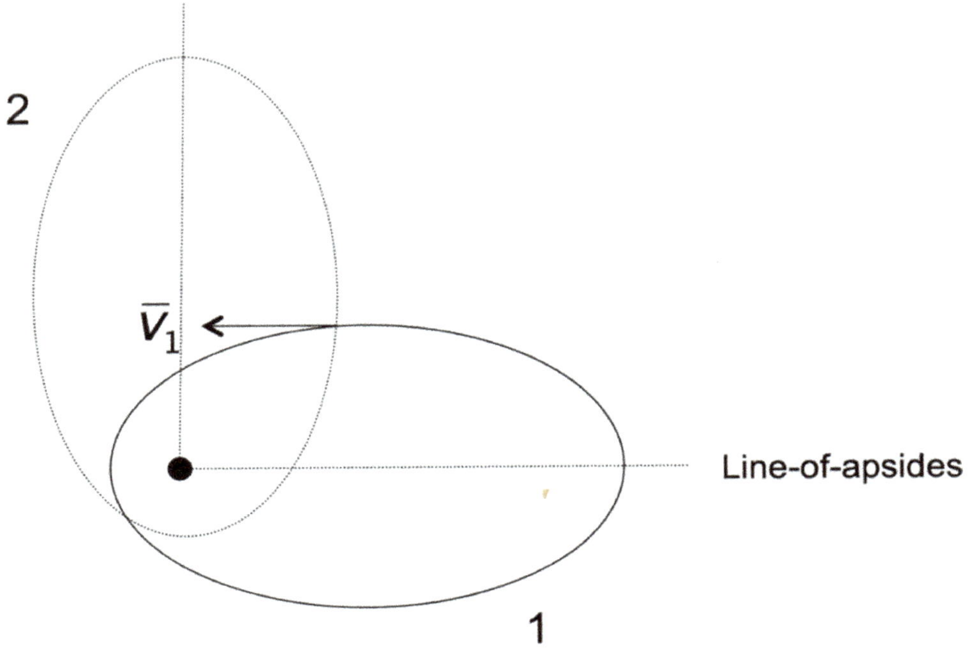

The velocities at the intersection of orbit 1 and orbit 2 are equal in magnitude but have different directions. A $\Delta\bar{v}$ must be applied at the intersection to rotate $\bar{v}_1$ to orbit 2 as

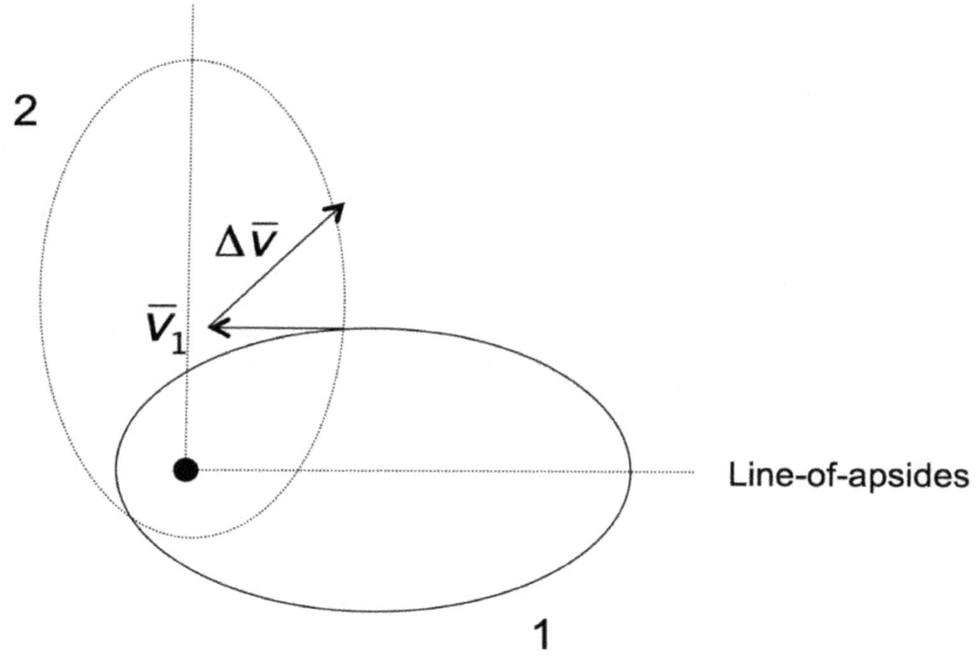


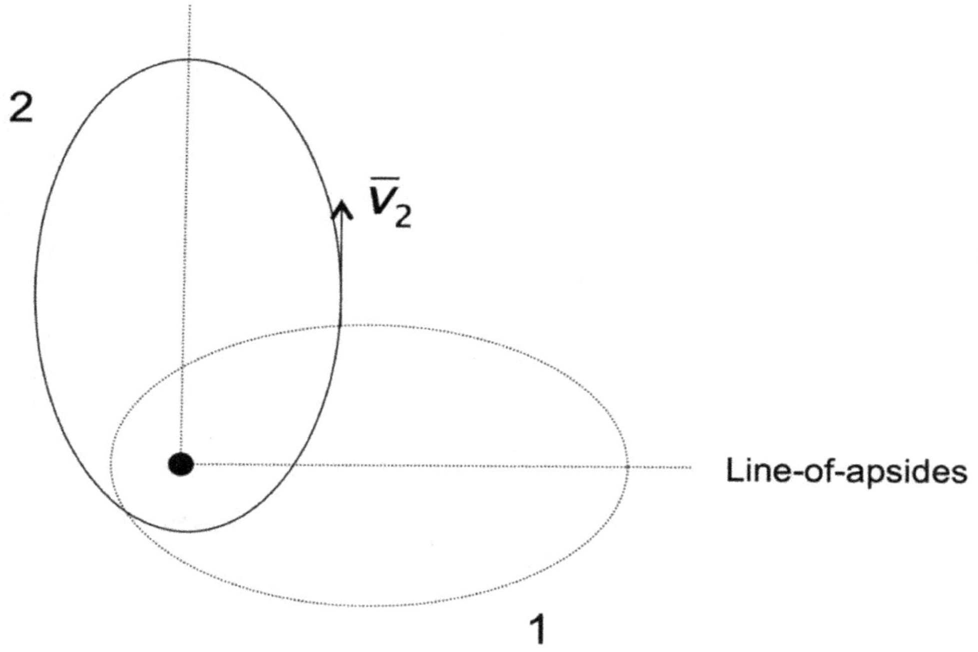

Example 15.3

For the $90^o$ rotation of the apse line shown above, calculate the $\Delta v$ required in the case where $a = 8{,}000$ km and $e = 0.2$ for an Earth orbit.

Solution:

$$r = \frac{a(1-e^2)}{1+e\cos\nu} = \frac{8{,}000\left[1-(0.2)^2\right]}{1+(0.2)\cos 45^o} = 6{,}728.5 \text{ km}$$

$$p = 7{,}680 \text{ km}$$

$$h = \sqrt{\mu p} = \sqrt{(3.986 \times 10^5)(7{,}680)} = 55{,}328.5 \ \frac{\text{km}^2}{\text{s}}$$

$$v = \sqrt{\mu\left(\frac{2}{r} - \frac{1}{a}\right)} = \sqrt{(3.986 \times 10^5)\left(\frac{2}{6{,}728.5} - \frac{1}{8{,}000}\right)} = 8.29 \ \frac{\text{km}}{\text{s}}$$

$$\varphi = \cos^{-1}\left(\frac{h}{rv}\right) = \cos^{-1}\left[\frac{55{,}328.5}{6{,}728.45(8.29)}\right] = 7.29^o$$

$$\Delta v = 2v\sin\varphi = 2(8.29)\sin 7.29^o$$

$$\Rightarrow \quad \Delta v = 2.10 \ \frac{\text{km}}{\text{s}}$$

## Problems

15.1 A spacecraft is traveling in an elliptical Earth orbit having a semi-major axis of 11,500 km and an eccentricity of 0.3. Find the magnitude of a single-impulse maneuver which will rotate the perigee of the orbit by $180^o$ without changing the size or shape of the orbit.
(Ans. $\Delta v = 3.68 \ \frac{\text{km}}{\text{s}}$)

15.2 For the maneuver described in Problem 15.1, express the velocity change in terms of the eccentricity e and the angular momentum h.
(Ans. $\Delta v = \frac{2e\mu}{h}$)

15.3 For the spacecraft depicted in Problem 15.1, calculate the magnitude of a single impulse maneuver which will rotate the perigee of the orbit by $90^o$ without changing the size or shape of the orbit.
(Ans. $\Delta v = 2.57 \ \frac{\text{km}}{\text{s}}$)

15.4 For the maneuver described in Problem 15.3, express the velocity change in terms of the eccentricity e and the angular momentum h.

(Ans. $\Delta v \frac{\sqrt{2}\mu e}{h}$)

15.5 A satellite in an Earth orbit has $a = 12{,}000$ km and $e = 0.15$. Calculate the magnitude of the $\Delta v$ required at perigee to rotate the apse line $90^o$ clockwise and reduce the semi-major axis while keeping the same eccentricity. At the point of application of the velocity change $r = p$ on the new orbit. Also find the semi-major axis of the new orbit.

(Ans. $\Delta v = 1.04\ \frac{km}{s}$, $a = 10{,}434.8$ km)

15.6 For Problem 15.5, express the magnitude of the angular momentum of the new orbit, $h_2$, in terms of that for the original orbit., $h_1$.

(Ans. $h_2 = \frac{h_1}{\sqrt{1+e}}$)

15.7 For Problem 15.5, find the semi-major axis of the new orbit and the $\Delta v$ required for the case where the eccentricity is increased to 0.3.

(Ans. $a = 11{,}208.8$ km, $\Delta v = 1.76\ \frac{km}{s}$)

15.8 For Problem 15.7, express the semi-major axis of the new orbit, $a_2$, in terms of the semi-major axis for the original orbit, $a_2$ .

(Ans. $a_2 = a_1 \left[\frac{1-e_1}{1-e_2^2}\right]$)

# Module 16: Single-Impulse Maneuvers - Intercept and Rendezvous

Important and useful applications of a single-impulse maneuver are those of intercept and rendezvous, which are discussed in this module. An intercept implies the chase vehicle colliding with a target, whereas a rendezvous implies the chase vehicle matching velocities with a target for the purpose of docking or capture. In true rendezvous problems, the initial velocity change ensures the vehicle will reach the target, but a second velocity change is usually required to match the target's velocity upon arrival.

In any type of intercept or rendezvous problem, the most important part is to ensure the target vehicle is in the proper location when the chase vehicle arrives. This is accomplished by departing on the intercept trajectory when the target is in the proper position, which requires the appropriate phase angle at departure. Calculation of this angle is discussed in later modules.

## Intercept Problems

Consider that a spacecraft in the circular orbit 1 having velocity $\bar{v}_c$ needs to intercept a spacecraft in the circular orbit 2

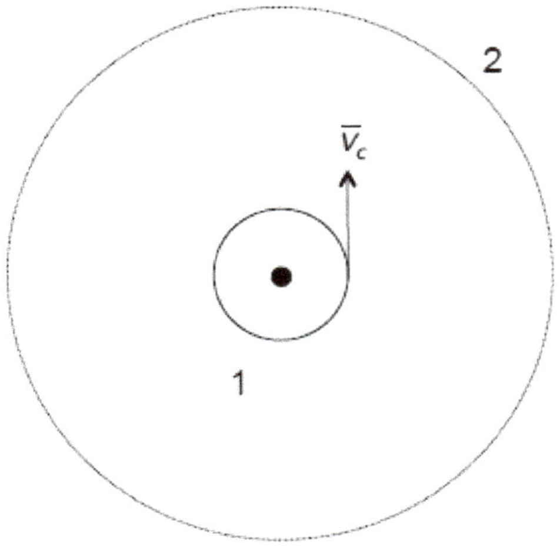

A tangential $\Delta\bar{v}$ in the direction of travel increases the velocity in order to reach orbit 2

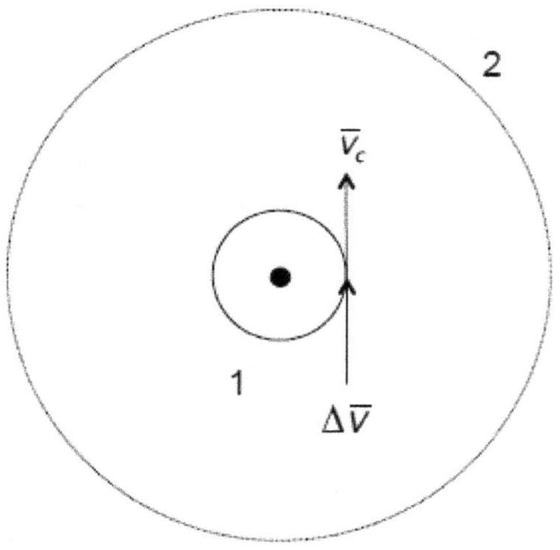


The amount of $\Delta\bar{v}$ applied will determine the path from orbit 1 to orbit 2. A larger $\Delta\bar{v}$ will create a faster transfer but will use more fuel, and a smaller $\Delta\bar{v}$ takes more time but uses less fuel, as shown in this figure.

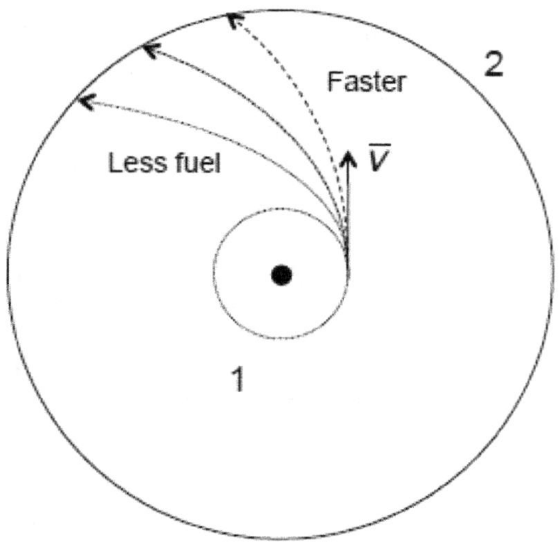

Example 16.1

Considering the figure shown above, calculate the $\Delta v$ required for the intercept trajectories from orbit 1 to orbit 2 having eccentricities of 1.0, 0.75, 0.50, 0.25, and 0.1. Express $\Delta v$ in terms of the circular orbit velocity, $v_c$ for each case.

Solution:

$$v_c = \sqrt{\frac{\mu}{a_1}}$$

$e = 1: \implies a_T = \infty$

$$v_p = \sqrt{\mu\left(\frac{2}{a_1} - \frac{1}{a_T}\right)} = \sqrt{\mu\left(\frac{2}{a_1} - \frac{1}{\infty}\right)} = \sqrt{\frac{2\mu}{a_1}} = 1.41v_c$$

$$\Delta v = 1.41v_c - v_c$$

$$\implies \Delta v = 0.41v_c$$

$e = 0.75:$

$$a_1 = a_T(1-e) = 0.25a_T \implies a_T = 4a_1$$

$$v_p = \sqrt{\mu\left(\frac{2}{a_1} - \frac{1}{a_T}\right)} = \sqrt{\mu\left(\frac{2}{a_1} - \frac{1}{4a_1}\right)} = \sqrt{\frac{7\mu}{4a_1}} = 1.32v_c$$

$$\Delta v = 1.32v_c - v_c$$

$$\implies \Delta v = 0.32v_c$$

$e = 0.50:$

$$a_1 = a_T(1-e) = 0.50a_T \implies a_T = 2a_1$$

$$v_p = \sqrt{\mu\left(\frac{2}{a_1} - \frac{1}{a_T}\right)} = \sqrt{\mu\left(\frac{2}{a_1} - \frac{1}{2a_1}\right)} = \sqrt{\frac{3\mu}{2a_1}} = 1.22v_c$$

$$\Delta v = 1.22v_c - v_c$$

$$\implies \Delta v = 0.22v_c$$

$e = 0.25$:

$$a_1 = a_T(1 - e) = 0.75a_T \Rightarrow \; a_T = 1.33a_1$$

$$v_p = \sqrt{\mu\left(\frac{2}{a_1} - \frac{1}{a_T}\right)} = \sqrt{\mu\left(\frac{2}{a_1} - \frac{1}{1.33a_1}\right)} = \sqrt{\frac{1.66\mu}{1.33a_1}} = 1.12v_c$$

$$\Delta v = 1.12v_c - v_c$$

$$\Rightarrow \; \Delta v = 0.12v_c$$

$e = 0.1$:

$$a_1 = a_T(1 - e) = 0.9a_T \Rightarrow \; a_T = 1.11$$

$$v_p = \sqrt{\mu\left(\frac{2}{a_1} - \frac{1}{a_T}\right)} = \sqrt{\mu\left(\frac{2}{a_1} - \frac{1}{1.11a_1}\right)} = \sqrt{\frac{1.22\mu}{1.11a_1}} = 1.05v_c$$

$$\Delta v = 1.05v_c - v_c$$

$$\Rightarrow \; \Delta v = 0.05v_c$$

Example 16.2

A space-based interceptor is parked in a circular Earth orbit at an altitude of 500 km and must perform a fast intercept of a target located in another circular orbit at an altitude of 800 km. A rocket firing occurs placing the interceptor on a trajectory to hit the target at a point where the true anomaly of the intercept trajectory is $110^o$. Determine the magnitude of the velocity change required for this intercept.

Solution:

$$r_1 = R_E + (\text{alt})_1 = 6{,}378.1 + 500 = 6{,}878.1 \text{ km}$$

$$v_c = \sqrt{\frac{\mu}{r_1}} = \sqrt{\frac{3.986\times10^5}{6{,}878.1}} = 7.61 \; \frac{\text{km}}{\text{s}}$$

$$r_2 = R_E + (\text{alt})_2 = 6{,}378.1 + 800 = 7{,}178.1 \text{ km}$$

$$r_p = a_2(1 - e_2)$$

$$\Rightarrow \; 6{,}878.1 = a_2(1 - e_2)$$

$$r_2 = \frac{a_2(1-e_2^2)}{1+e_2\cos\nu} = \frac{a_2(1-e_2)(1+e_2)}{1+e_2\cos\nu} = \frac{r_p(1+e_2)}{1+e_2\cos\nu}$$

$$7{,}178.1 = \frac{6{,}878.1(1+e_2)}{1+e_2\cos 110^o}$$

$$7{,}178.1(1-0.342e_2) = 6{,}878.1(1+e_2)$$

$$7{,}178.1 - 2{,}454.9e_2 = 6{,}878.1 + 6{,}878.1e_2$$

$$\Rightarrow \quad e_2 = 0.032$$

$$6{,}878.1 = a_2(1-e_2) = a_2(1-0.032)$$

$$\Rightarrow \quad a_2 = 7{,}106.5 \text{ km}$$

$$v_p = \sqrt{\mu\left(\frac{2}{r_p} - \frac{1}{a_2}\right)} = \sqrt{3.986 \times 10^5\left(\frac{2}{6{,}878.1} - \frac{1}{7{,}106.5}\right)}$$

$$v_p = 7.73\ \frac{\text{km}}{\text{s}}$$

$$\Delta v = v_p - v_c = 7.73 - 7.61\ \frac{\text{km}}{\text{s}}$$

$$\Rightarrow \quad \Delta v = 0.12\ \frac{\text{km}}{\text{s}}$$

Rendezvous Problems

Although rendezvous problems typically consist of two or more velocity changes, they'll be introduced in this module to demonstrate their contrast with intercept problems.

Chase maneuvers are used to achieve a rendezvous with another body in space. Such a chase maneuver is often referred to as Lambert's Problem, which can be stated as follows:

> Given the position vectors at two points on an ellipse and the true anomaly between them, calculate the velocity vectors at both points.

The difference between intercept and rendezvous problems is that for rendezvous, the velocity of the chase vehicle at the rendezvous point must match that of the target so docking or other proximity operations can occur. Rendezvous trajectories near a planet are likely too fuel intensive to be practical and are therefore commonly performed at high altitudes.

Consider the problem where vehicle A, having velocity $\bar{v}_{A1}$, is to rendezvous with target B, having velocity $\bar{v}_B$, at point C. Points A and B are both located in orbit 1 as shown

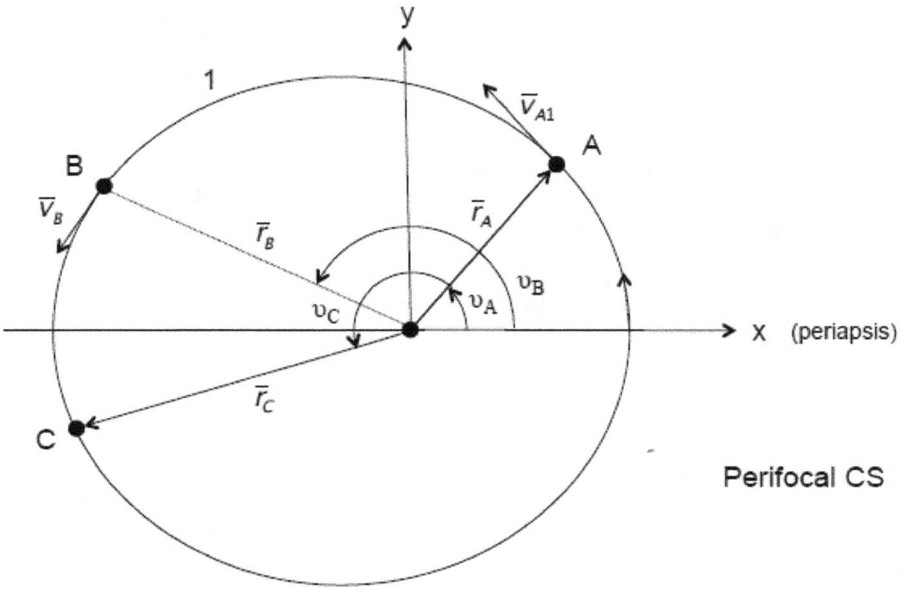


The rendezvous problem can be stated as

- Given: $a_1$, $e_1$, $\nu_A$ , $\nu_B$, $\nu_C$ and the $\Delta t_{AC}$ to intercept.
- Calculate: $a_2$, $e_2$, of an intercept trajectory and the $\Delta v_A$ required to achieve intercept at point C.
- Calculate: $\Delta v_C$ to match $\bar{v}_{C1}$ for rendezvous.

This problem answers the question

"How does a spacecraft get from point A to point B in a given amount of time?"

The answer determines the nature of the maneuver required.

Vehicle A will travel to point C along orbit 2 below. A velocity change $\bar{v}_A$ is applied at point A to create the velocity required to enter orbit 2, $\bar{v}_{A2}$ as shown

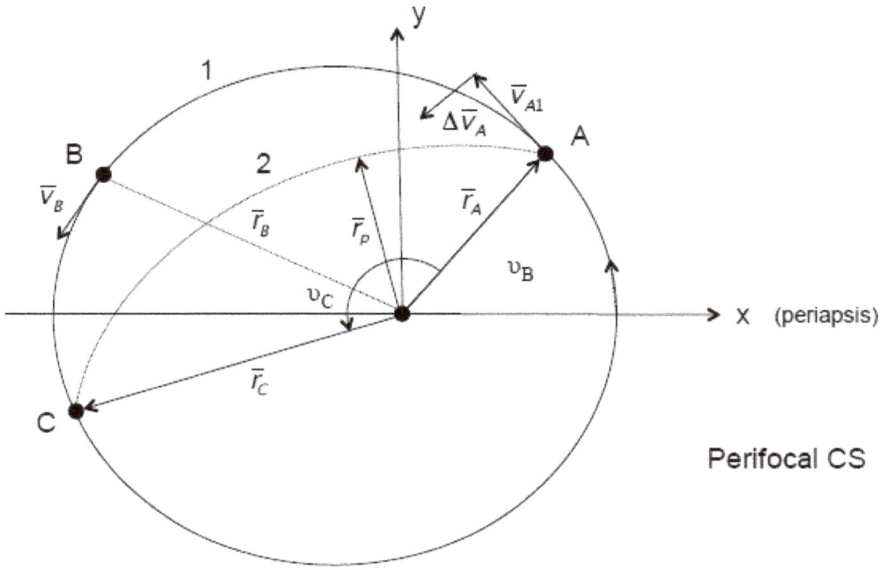


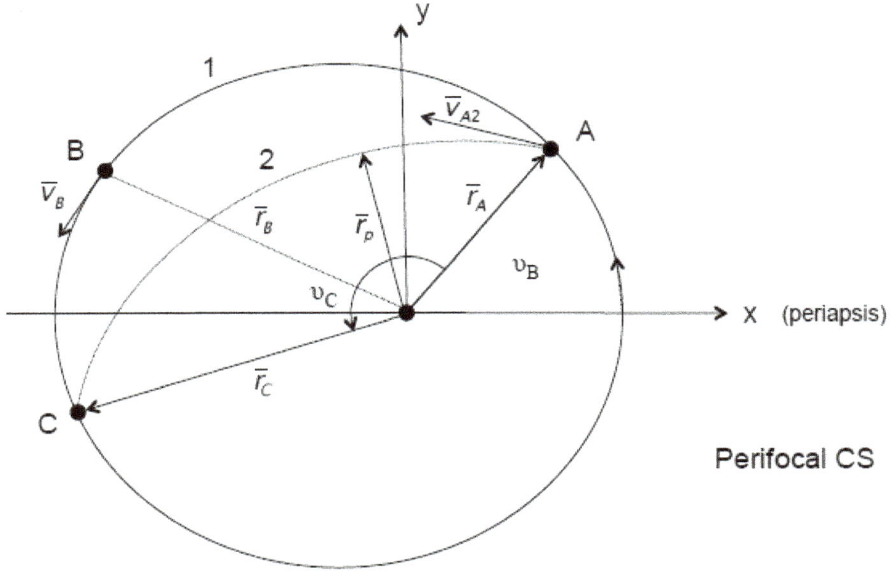

Vehicle A will reach point C with velocity $\bar{v}_{C2}$

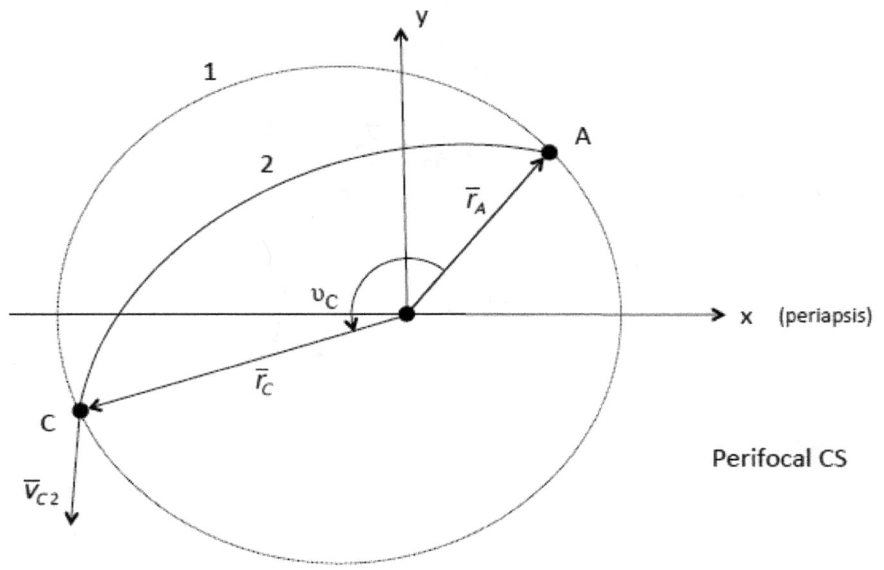


A second velocity change is required, $\Delta\bar{v}_C$, to match the velocity to maintain orbit 1 at point C, $\bar{v}_{C1}$ , as shown

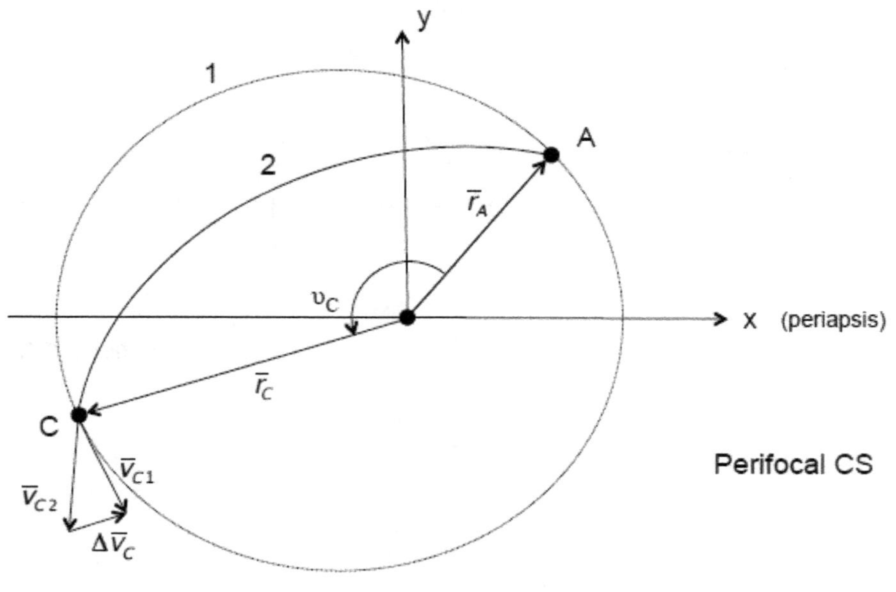

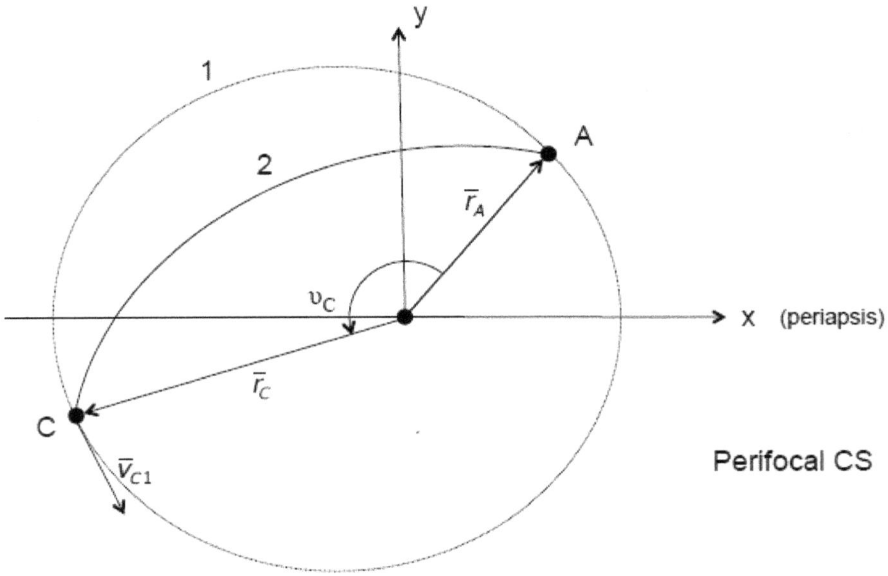


The solι tion to the rendezvous problem can be stated as

- Calculate $r_A, \bar{v}_{A1}, r_C$, and $\bar{v}_{C1}$ by standard two-body methods.
- Calculate the required $\Delta\bar{v}_A$ using Lambert's Theorem, given by

$$\bar{v}_{A2} = \left[\frac{\sqrt{\mu p}}{r_A r_C \sin \nu_C}\right]\left\{\bar{r}_C - \left[\frac{1 - r_C(1 - \cos \nu_C)}{p}\right]\bar{r}_A\right\}$$

$$\Delta\bar{v}_A = \bar{v}_{A2} - \bar{v}_{A1}$$

- Calculate the required $\Delta\bar{v}_C$ to match $\bar{v}_{C1}$ for rendezvous, again using Lambert's Theorem

$$\bar{v}_{C2} = \left[\frac{\sqrt{\mu p}}{r_A r_C \sin \nu_C}\right]\left\{\left[\frac{1 - r_C(1 - \cos \nu_C)}{p}\right]\bar{r}_C - \bar{r}_A\right\}$$

$$\Delta\bar{v}_C = \bar{v}_{C1} - \bar{v}_{C2}$$

- The magnitude of the total $\Delta\bar{v}$ required for rendezvous is calculated by

$$\Delta\bar{v}_{TOT} = |\bar{v}_A| + |\bar{v}_C|$$

Example 16.3

For the case of $e = 0.75$ in Example 16.1, determine the value of $a_2$ (in terms of $a_1$) for which the speed of the interceptor is equal to the speed of the target when it reaches the target.

Solution:

$$v_P = \sqrt{\mu\left(\frac{2}{a_2} - \frac{1}{a_T}\right)}$$

$$v_T = \sqrt{\frac{\mu}{a_2}}$$

$$v_P = v_T$$

$$\sqrt{\mu\left(\frac{2}{a_2} - \frac{1}{a_T}\right)} = \sqrt{\frac{\mu}{a_2}}$$

$$\sqrt{\mu\left(\frac{2}{a_2} - \frac{1}{4a_1}\right)} = \sqrt{\frac{\mu}{a_2}}$$

$$\frac{2}{a_2} - \frac{1}{4a_1} = \frac{1}{a_2}$$

$$\frac{1}{a_2} = \frac{1}{4a_1}$$

$$\Rightarrow \quad a_2 = 4a_1$$

Problems

16.1 A space vehicle is at an altitude of 500 km in a circular Earth orbit. Calculate the $\Delta v$ required for this vehicle to intercept another vehicle at an altitude of 1,000 km at a position 270° from the point of $\Delta v$ application. Also determine the speed of the space vehicle when it reaches its target.
(Ans. $\Delta v = 0.22\ \frac{km}{s}, v_2 = 7.37\ \frac{km}{s}$)

16.2 For Example 16.1, derive an expression for $\Delta v$ in terms of the eccentricity of the intercept trajectory.
(Ans. $\Delta v = \left[\sqrt{1+e} - 1\right]\sqrt{\frac{\mu}{a_1}}$)

16.3 A space-based interceptor is parked in a circular Earth orbit at an altitude of 400 km and must perform a fast intercept of a target located in another circular orbit at an altitude of 4,000 km. A rocket firing occurs placing the interceptor on a trajectory to hit the target at a point where the true anomaly of the intercept trajectory is 110°. Calculate the Δv required for this intercept.
(Ans. $\Delta v = 1.36\ \frac{km}{s}$)

16.4 A space-vehicle armed with an interceptor is parked in a circular Earth orbit at an altitude of 1,200 km. The target is traveling in another circular orbit at an altitude of 350 km and must be intercepted at a point having a true anomaly of 80°. Determine the Δv required to achieve intercept.
(Ans. $\Delta v = -0.36\ \frac{km}{s}$)

16.5 A satellite in a circular Earth orbit with semi-major axis $a_1$ is to rendezvous with another satellite in a circular Earth orbit with semi-major axis $a_2$, where $a_2 = 10a_1$. The satellite will transfer from $a_1$ to $a_2$ using an elliptical transfer ellipse and the rendezvous will occur when the true anomaly is equal to 180°. Find an expression for the Δv required in terms of the circular speed in the lower orbit.
(Ans. $\Delta v = 0.35\sqrt{\frac{\mu}{a_1}}$)

16.6 Solve Problem 16.5 for the case where the satellite begins in the higher orbit and will rendezvous the second other satellite in the lower orbit when the true anomaly is also equal to 180°. Express Δv in terms of the circular speed of the lower orbit.
(Ans. $\Delta v = -0.18\sqrt{\frac{\mu}{a_1}}$)

16.7 Consider a satellite in a circular Earth orbit where r = 8,000 km. In order to rendezvous with a space vehicle ahead in the same orbit, the satellite must perform a fast transfer covering a span of 90° of the circular orbit using an elliptical orbit having $r_p = 7{,}000$ km. Calculate the magnitude of the velocity change necessary to complete this maneuver.
(Ans. $\Delta v = 3.01\ \frac{km}{s}$)

16.8 For the transfer described in Problem 16.7, find the magnitude of the Δv if a hyperbolic trajectory having e = 1.1 is used to increase the transfer speed to reach the rendezvous position.
(Ans. $\Delta v = 3.97\ \frac{km}{s}$)

(This page was intentionally left blank.)

# Module 17: Hohmann and Bielliptic Transfers

## Hohmann Transfers

Hohmann transfers are transfers between two circular, coplanar orbits of different sizes ($a_1$and $a_2$) using an elliptical transfer orbit with two tangential impulses, one at each end of the transfer ellipse.

Hohmann transfers are one of the most useful types of maneuvers for inserting satellites into high altitude orbits. They are minimum energy transfers if $\frac{a_2}{a_1} \leq 11.94$, otherwise a bi-elliptic transfer may be more efficient.

Consider a transfer from circular orbit 1 to a higher altitude circular orbit 2 using a Hohmann transfer. The circular velocity in orbit 1 is $\bar{v}_{c1}$.

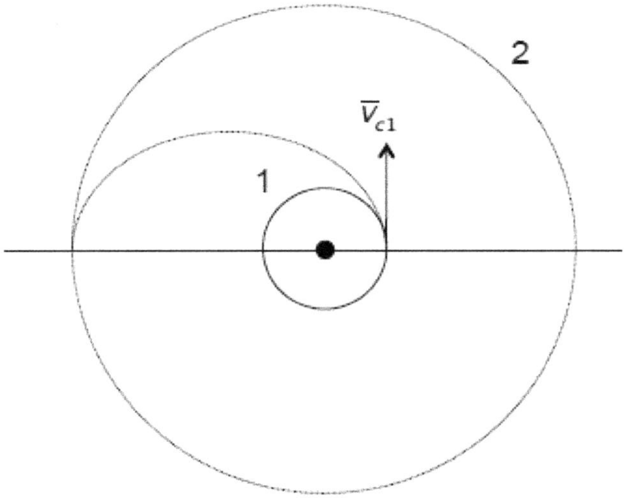

A tangential velocity change $\Delta\bar{v}_1$ in the direction of travel increases the velocity to move to the transfer ellipse

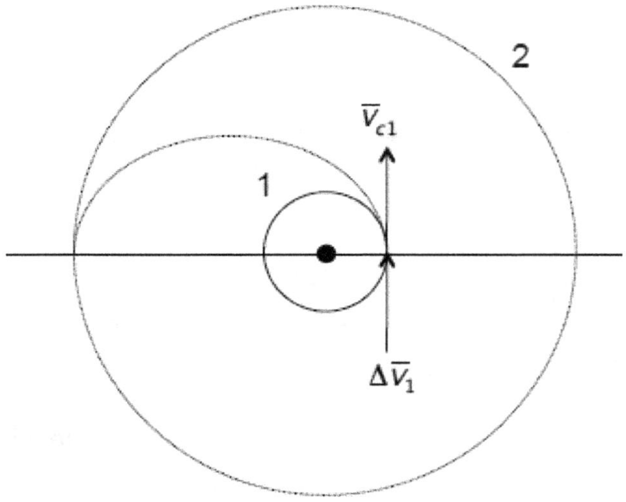


The speed at the periapsis of the transfer ellipse will be $\bar{v}_{T1}$

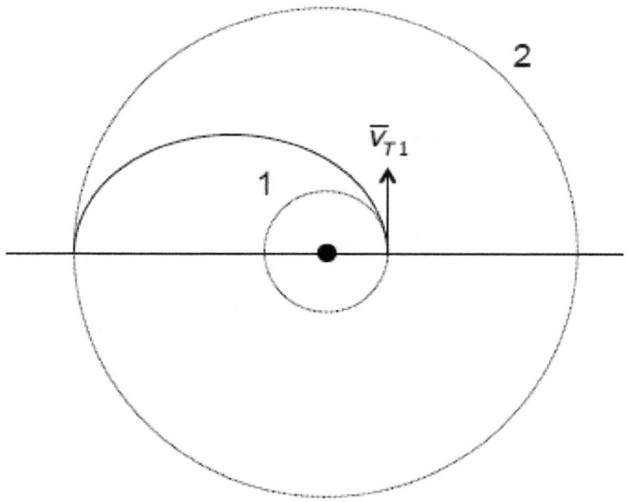

As the spacecraft travels along the transfer ellipse its velocity will decrease since as it moves farther away from the primary mass. When it reaches the apoapsis of the transfer ellipse the velocity will be $\bar{v}_{T2}$.

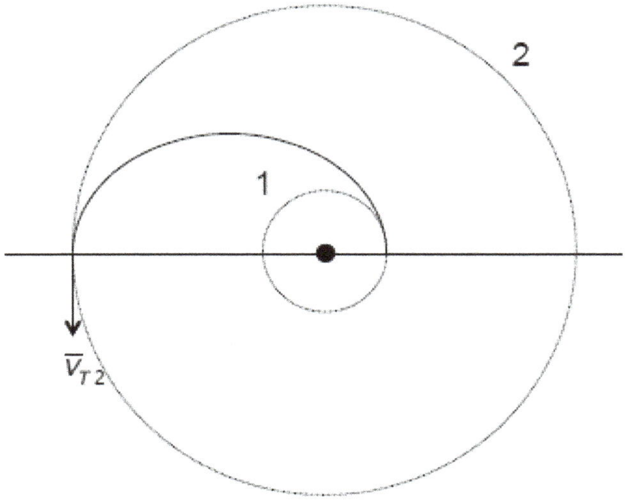


At the apoapsis of the transfer ellipse a second tangential $\Delta\bar{v}_2$ in the direction of travel is required to increase the velocity to maintain the circular orbit 2, $\bar{v}_{c2}$.

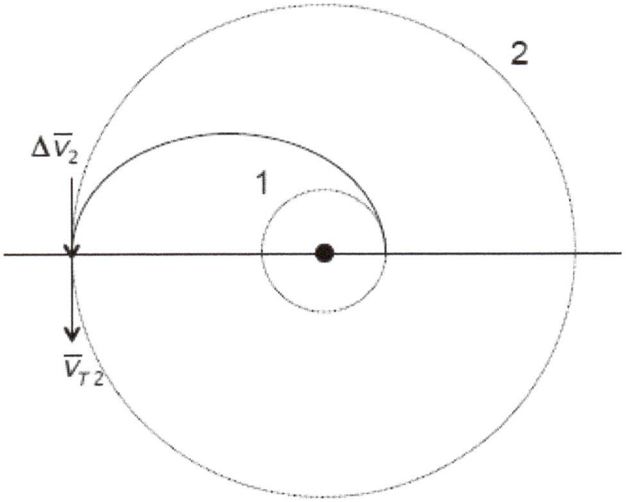

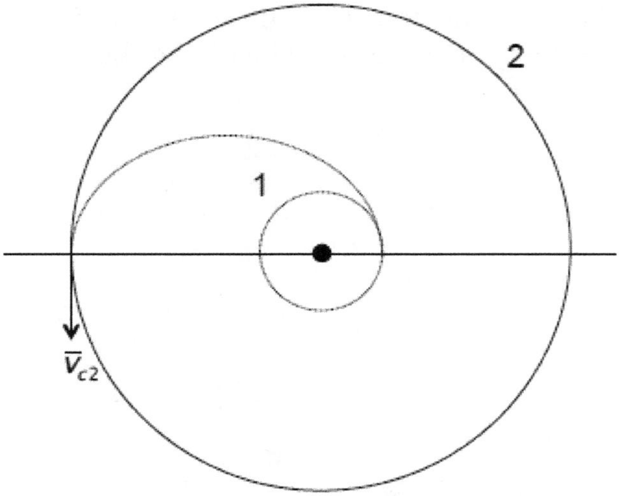


The calculations for the magnitudes of the velocities required for a Hohmann transfer are given as

$$v_{c1} = \sqrt{\frac{\mu}{a_1}}$$

$$v_{c2} = \sqrt{\frac{\mu}{a_2}}$$

$$a_T = \frac{a_1 + a_2}{2}$$

$$v_{T1} = \mu\sqrt{\frac{2}{a_1} - \frac{1}{a_T}}$$

$$v_{T2} = \mu\sqrt{\frac{2}{a_2} - \frac{1}{a_T}}$$

$$\Delta v_1 = v_{T1} - v_{c1}$$

$$\Delta v_2 = v_{c2} - v_{T2}$$

$$\Delta v = \Delta v_1 + \Delta v_2$$

where

$a_1$ = semi-major axis of orbit 1
$a_2$ = semi-major axis of orbit 2

For intercept and rendezvous problems, phasing is the most important aspect. In order for the target to be in the appropriate location when $m_2$ arrives, the target must be in the proper location when $m_2$ departs.

Consider the case where $m_2$ departs orbit 1 to intercept or rendezvous with a target in orbit 2. Departure must occur at the proper phase angle at departure, γ, to ensure the target will be in the appropriate location when $m_1$ arrives at orbit 2

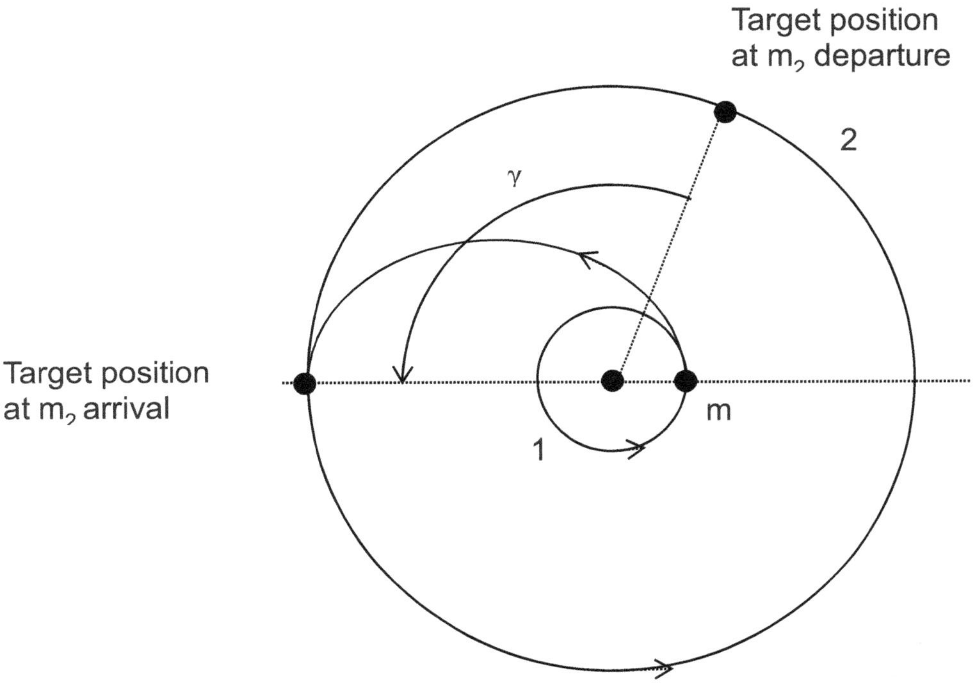


For a Hohmann transfer between two circular orbits, the phase angle at departure is determined by the equation

$$\gamma = 180\sqrt{\left(\frac{a_1+a_2}{2a_2}\right)^3}$$

For other types of intercept or rendezvous trajectories, when the target vehicle is in a circular orbit, the phase angle at departure can be calculated using the expression

$$\gamma = (360)\frac{t_T}{T}$$

where $t_T$ is the transfer time to the target.

It's also common practice to perform Hohmann-type transfers between two coplanar, elliptical orbits, as

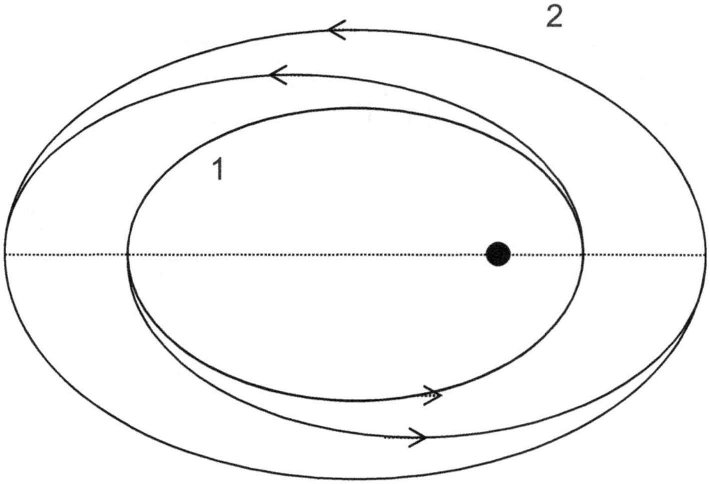


Departure can take place at either the periapsis or apoapsis of orbit 1 resulting in arrival at either the apoapsis or periapsis of orbit 2. Velocity changes required in these types of Hohmann transfers are calculated in a similar manner as discussed above.

Example17.1

Calculate the total $\Delta v$ necessary to move an Earth satellite from one circular orbit with an altitude of 500 km to another circular orbit with an altitude of 2,000 km having the same inclination using a Hohmann transfer.

Solution:

$$r_1 = R_E + (\text{alt})_1 = 6{,}378.1 + 500 = 6{,}878.1 \text{ km}$$

$$r_2 = R_E + (\text{alt})_2 = 6{,}378.1 + 2{,}000 = 8{,}378.1 \text{ km}$$

$$a_T = \frac{(a_1 + a_2)}{2} = \frac{(6{,}878.1 + 8{,}378.1)}{2} = 7{,}628.1 \text{ km}$$

$$v_{c1} = \sqrt{\frac{\mu}{a_1}} = \sqrt{\frac{3.986\times10^5}{6{,}878.1}} = 7.61 \frac{\text{km}}{\text{s}}$$

$$v_{T1} = \sqrt{\mu\left(\frac{2}{r_1} - \frac{1}{a_T}\right)} = \sqrt{\mu\left(\frac{2}{6{,}878.1} - \frac{1}{(7{,}628.1)}\right)} = 7.98 \frac{\text{km}}{\text{s}}$$

$$v_{T2} = \sqrt{\mu\left(\frac{2}{r_2} - \frac{1}{a_T}\right)} = \sqrt{\mu\left(\frac{2}{8{,}378.1} - \frac{1}{(7{,}628.1)}\right)} = 6.55\ \frac{km}{s}$$

$$v_{c2} = \sqrt{\frac{\mu}{a_2}} = \sqrt{\frac{3.986\times10^5}{8{,}378.1}} = 6.90\ \frac{km}{s}$$

$$\Delta v = (v_{T1} - v_{c1}) + (v_{c2} - v_{T2}) = (7.98 - 7.61) + (6.90 - 6.55)$$

$$\Rightarrow\ \Delta v = 0.72\ \frac{km}{s}$$

Example 17.2

A spacecraft in a circular orbit where $r = a_1$, is to rendezvous with another spacecraft in a circular orbit where $r = a_2 > a_1$ using a Hohmann transfer. Show that the phase angle at departure, $\gamma$ (in degrees) can be expressed as

$$\gamma = \sqrt{\frac{4{,}050(a_1+a_2)^3}{a_2^3}}$$

Solution:

$$a_T = \frac{(a_1+a_2)}{2}$$

$$t_T = \frac{T_T}{2} = \frac{2\pi}{2}\sqrt{\frac{a_T^3}{\mu}} = \pi\sqrt{\frac{(a_1+a_2)^3}{8\mu}}$$

$$T_T = 2\pi\sqrt{\frac{a_2^3}{\mu}}$$

$$\gamma = \frac{t_T}{T_2}(360) = \frac{\pi\sqrt{\frac{(a_1+a_2)^3}{8\mu}}}{2\pi\sqrt{\frac{a_2^3}{\mu}}}(360) = \frac{\pi}{2\pi}\sqrt{\frac{(a_1+a_2)^3}{8\mu}\cdot\frac{\mu}{a_2^3}}(360) = \sqrt{\frac{(180)^2}{8}\cdot\frac{(a_1+a_2)^3}{a_2^3}}$$

$$\Rightarrow\ \gamma = \sqrt{\frac{4{,}050(a_1+a_2)^3}{a_2^3}}$$

<u>Bi-Elliptic Transfers</u>

For the case of $\frac{a_2}{a_1} > 11.94$, a bi-elliptic transfer may be more efficient than a Hohmann transfer.

A bi-elliptic transfer consists of transferring from $a_1$ to a distance far beyond $a_2$ before transferring back to $a_2$, then applying a final $\Delta v$ to maintain the higher altitude orbit. A bi-elliptic transfer is a three impulse maneuver as described below.

Beginning in circular orbit 1 with a circular velocity of $\bar{v}_{c_1}$, consider a transfer to a higher circular orbit 2

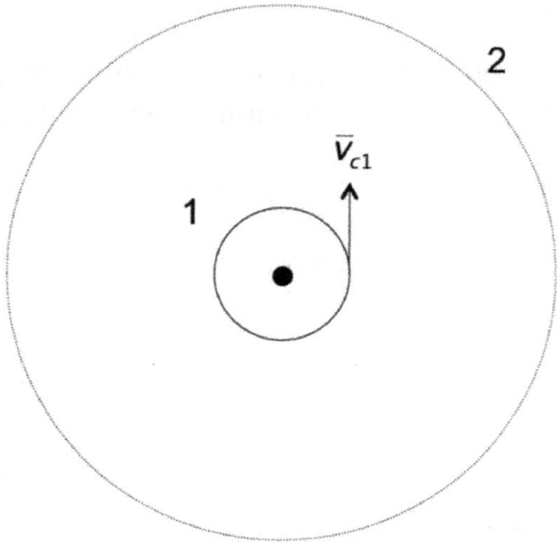


The first tangential $\Delta\bar{v}_1$ is applied in the direction of travel as shown

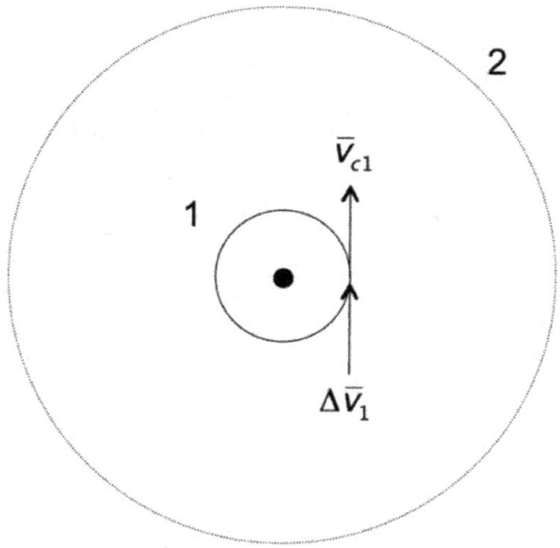

The velocity change is large enough to create transfer velocity $\bar{v}_{T1}$, which creates a transfer ellipse having an apoapsis far above orbit 2

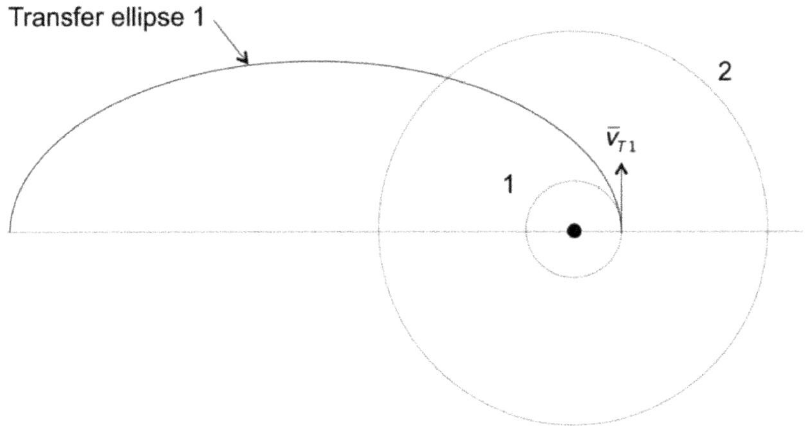


The velocity at the apoapsis of the first transfer ellipse will be $\bar{v}_{T2}$

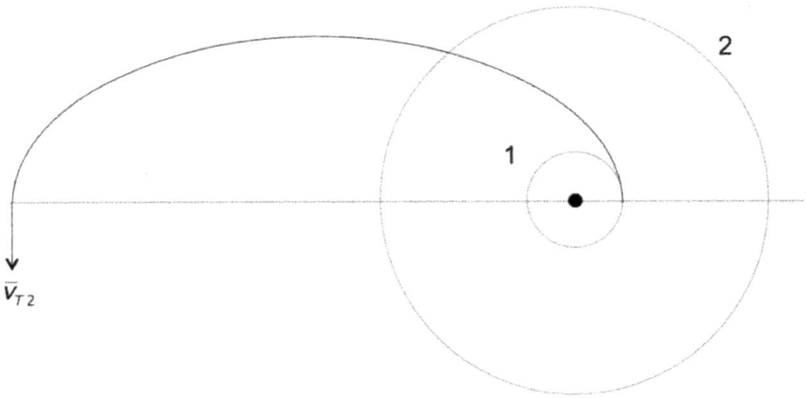

A second tangential $\Delta\bar{v}_2$ is then applied in the direction of the velocity to raise the periapsis of the transfer ellipse to match orbit 2

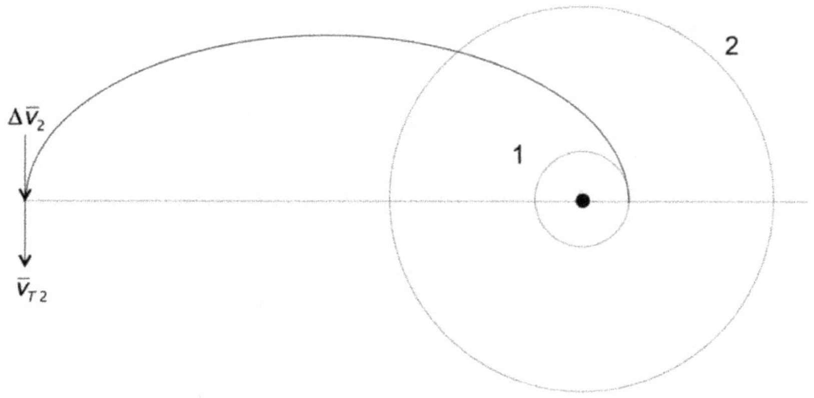


This creates the transfer velocity $\bar{v}_{T3}$ and the second transfer ellipse appears as

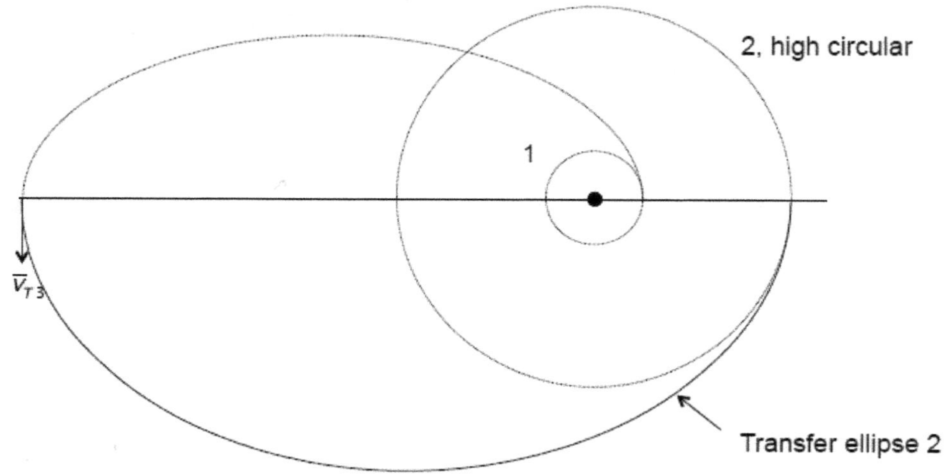

This creates the velocity $\bar{v}_{T4}$ at the periapsis of the second transfer ellipse, which coincides with the circular orbit 2

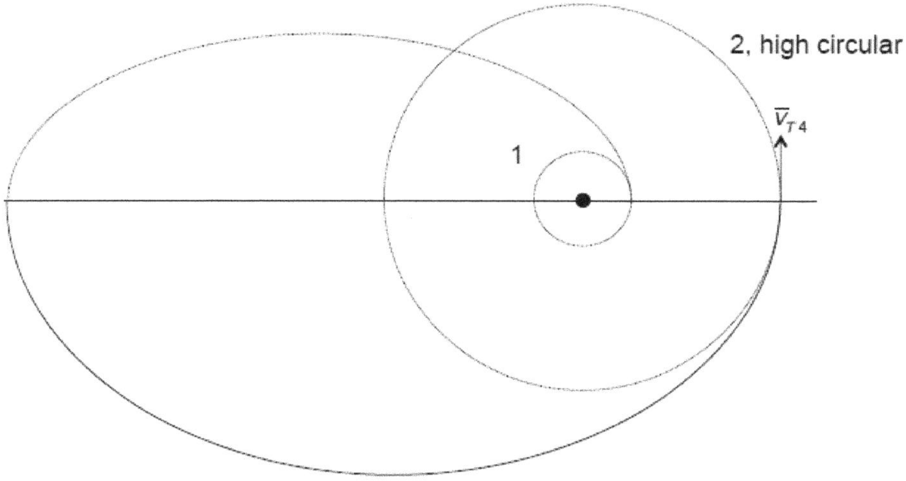


Finally, a third tangential velocity change $\Delta\bar{v}_3$ is applied in the opposite direction of the velocity to reduce the speed

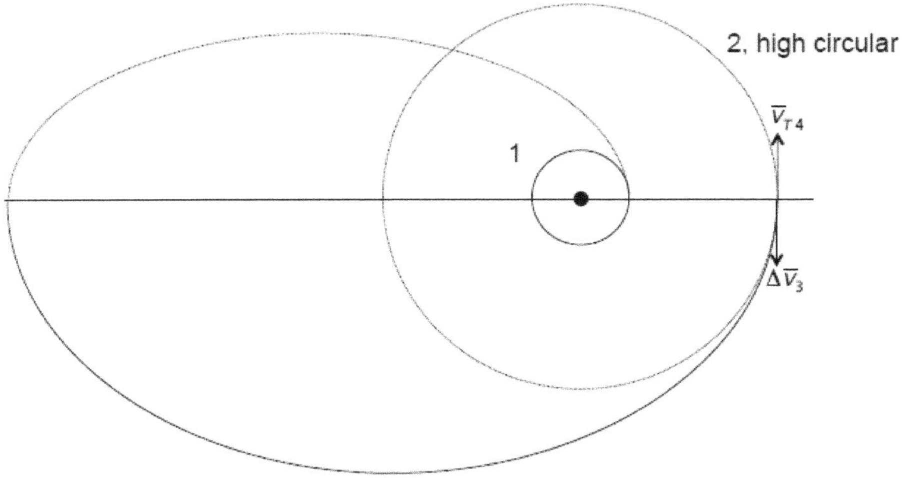

This velocity change creates velocity , $\bar{v}_{c2}$ which will maintain circular orbit 2

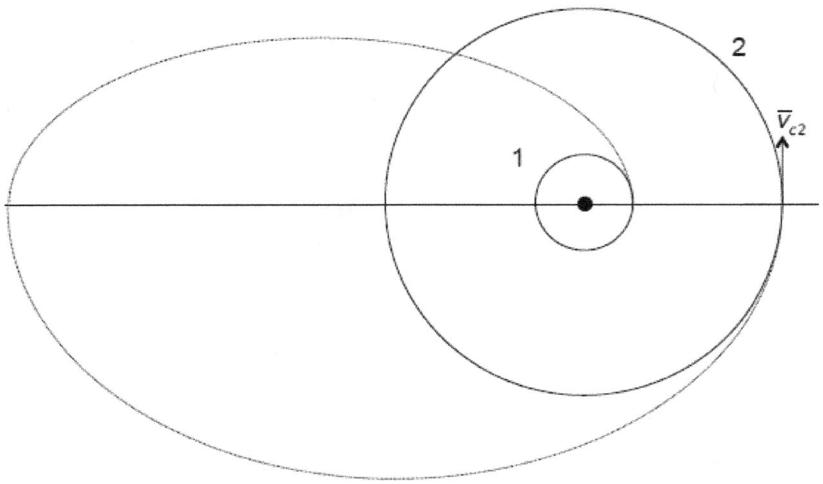


<u>Example 17.3</u>

Consider a bi-elliptic transfer from orbit 1 to orbit 2 where $a_2 = 15a_1$ and $r_{T2} = 2a_2$. Determine the total $\Delta v$ (in terms of $v_{c1}$required for a spacecraft to complete this transfer.

Solution:

$$v_{c1} = \sqrt{\frac{\mu}{a_1}}$$

$$v_{c2} = \sqrt{\frac{\mu}{a_2}} = \sqrt{\frac{\mu}{15a_1}} = 0.258\sqrt{\frac{\mu}{a_1}} = 0.258v_{c_1}$$

$$a_{T1} = \frac{a_1+2a_2}{2} = \frac{a_1+2(15a_1)}{2} = \frac{31a_1}{2} = 15.5a_1$$

$$a_{T2} = \frac{2a_2+a_2}{2} = \frac{3(15a_1)}{2} = 22.5a_1$$

$$v_{T1} = \sqrt{\mu\left(\frac{2}{a_1} - \frac{1}{a_{T_1}}\right)} = \sqrt{\mu\left(\frac{2}{a_1} - \frac{1}{15.5a_1}\right)} = 1.391\sqrt{\frac{\mu}{a_1}} = 1.391v_{c_1}$$

$$v_{T2} = \sqrt{\mu\left(\frac{2}{r_{T_2}} - \frac{1}{a_{T_1}}\right)} = \sqrt{\mu\left(\frac{2}{30a_1} - \frac{1}{15.5a_1}\right)} = 0.046\sqrt{\frac{\mu}{a_1}} = 0.046v_{c_1}$$

$$v_{T3} = \sqrt{\mu\left(\frac{2}{r_{T_2}} - \frac{1}{a_{T_2}}\right)} = \sqrt{\mu\left(\frac{2}{30a_1} - \frac{1}{22.5a_1}\right)} = 0.149\sqrt{\frac{\mu}{a_1}} = 0.149v_{c_1}$$

$$v_{T4} = \sqrt{\mu\left(\frac{2}{a_2} - \frac{1}{a_{T_2}}\right)} = \sqrt{\mu\left(\frac{2}{15a_1} - \frac{1}{22.5a_1}\right)} = 0.298\sqrt{\frac{\mu}{a_1}} = 0.298v_{c_1}$$

$$\Delta v = (v_{T1} - v_{c1}) - (v_{T3} - v_{T2}) + (v_{T4} - v_{c2})$$

$$\Delta v = (1.391v_{c1} - 1.0v_{c1}) + (0.149v_{c1} - 0.046v_{c1}) + (0.298v_{c1} - 0.258v_{c1})$$

$$\Rightarrow \Delta v = 0.534v_{c1}$$

<u>Problems</u>

17.1 A spacecraft is in a 500 km circular Earth orbit. Calculate the total Δv required for a Hohmann transfer to a 2,800 km coplanar circular Earth orbit and the transfer time.
(Ans. $\Delta v = 4.02\ \frac{km}{s}, t_T = 59.66$ min)

17.2 Find the time it takes to perform a Hohmann transfer from the orbit of Earth to the orbit of Mars. Assume the orbits to be circular and coplanar.
(Ans. $t_T = 258.8$ days)

17.3 A satellite in a circular 6,700 km geocentric orbit is to transfer to a larger circular orbit using a Hohmann transfer. If the first Δv applied is equal to 1.5 $\frac{km}{s}$, determine the second Δv and radius of the larger circular orbit.
(Ans. $\Delta v = 1.19\ \frac{km}{s}, r_2 = 16{,}676$ km)

17.4 A space vehicle in a circular orbit at an altitude of 500 km above Earth executes a Hohmann transfer to a 1,000 km altitude circular orbit. Calculate the total Δv required for this transfer.
(Ans. $\Delta v = 0.26\ \frac{km}{s}$)

17.5 Determine the total Δv required for a Hohmann transfer from a circular Earth orbit having a radius of 12r to a circular orbit having a radius of r.
(Ans. $\Delta v = -0.535\sqrt{\frac{\mu}{r}}$)

17.6 Find the total $\Delta v$ required to achieve a Hohmann transfer from a circular Earth orbit having a radius of 3r to another circular orbit having a radius of r. Express the velocity change in terms of $\sqrt{\frac{\mu}{r}}$.

(Ans. $\Delta v = 0.394\sqrt{\frac{\mu}{r}}$)

17.7 A spacecraft is in a circular heliocentric orbit with $a = 1.2 \times 10^8$ km and needs to discard a spent nuclear fuel cell by launching it toward the Sun.
(a) Determine the size of an elliptical transfer orbit such that the perihelion will be equal to the radius of the Sun.
(b) Calculate the single $\Delta v$ necessary to place the fuel cell in the transfer orbit.
(c) How long will it take the fuel cell to reach the Sun?
(Ans. (a) $a_T = 6.017 \times 10^7$km; (b) ; $v = 30.72\ \frac{km}{s}$; (c) The fuel cell will never reach the Sun! Why?)

17.8 Consider an Earth satellite in a 300 km circular orbit which will transfer to a 3,000 km altitude coplanar circular orbit using a bi-elliptic transfer. If the first transfer ellipse has an eccentricity of 0.3, determine the total $\Delta v$ required and the total transfer time.
(Ans. $\Delta v = 2.04\ \frac{km}{s}$, $t_T = 171.54$ min)

# Module 18: Plane Changes

Changing the Inclination of an Orbit

Transfer between two non-coplanar orbits having the same size and shape requires a rotation of the orbital plane through an angle equal to their difference in inclinations, as shown by

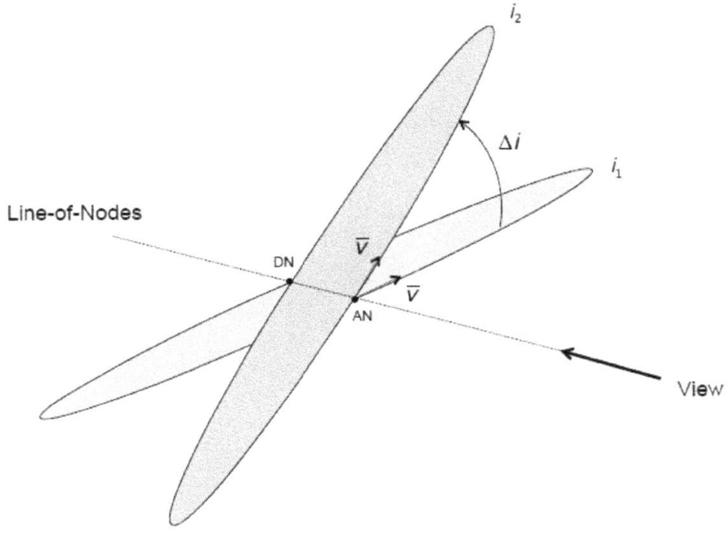


where

$$\Delta i = i_2 - i_1$$

The plane change is best accomplished by performing a rotation of the orbital plane about the line-of-nodes, i.e., orbital plane rotation, shown here at the ascending node

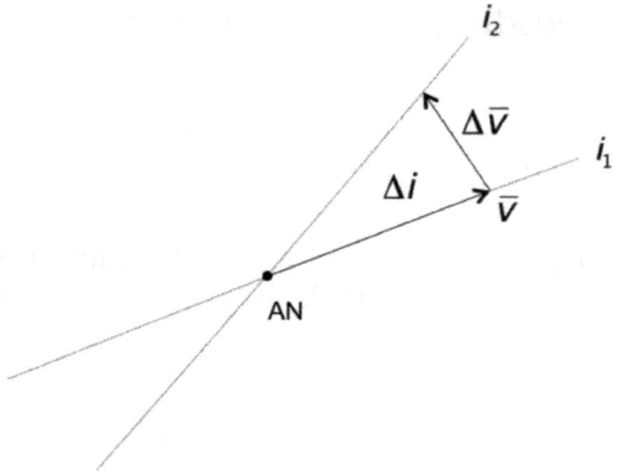


The rotation is achieved by applying a $\Delta\bar{v}$ in the direction shown

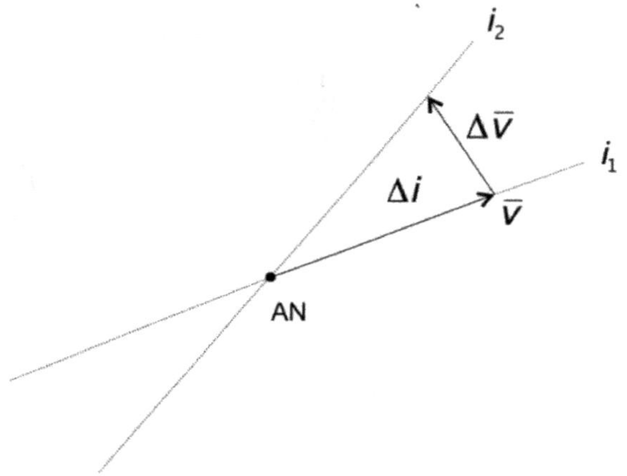


This velocity change will match the magnitude of the velocity in orbit 2, which is unchanged in magnitude from orbit 1 since the size and shape of the two orbits are identical

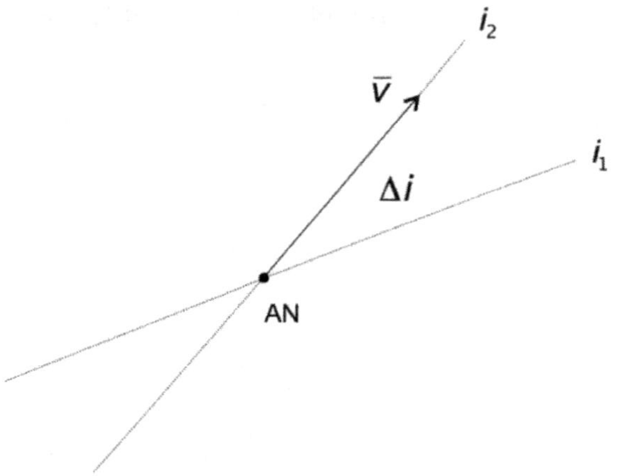

The magnitude of the required velocity change can be calculated by

$$\Delta v = 2v \sin\left(\frac{\Delta i}{2}\right)$$

Plane changes are very expensive in terms of fuel required. For example

$$\text{if } \Delta i = 60^o \implies \Delta v = v$$

Therefore, plane changes should be performed where $\bar{v}$ is the smallest, i.e., as close to apoapsis as possible. If this maneuver is performed at a node, it should be done at the node nearest the apoapsis.

A simple Rule of Thumb can be written as guidance regarding where to perform plane change maneuvers.

Rule of Thumb for Plane Changes:

1. Apply the $\Delta\bar{v}$ at the apoapsis, if the apoapsis is on the line-of-nodes, i.e.,

$$\omega = 0^o \text{ and } \upsilon = 180^o.$$

2. Otherwise, apply the $\Delta\bar{v}$ at the node closest to the apoapsis

$$\implies \text{at the DN, if } \omega < 90^o, \text{ or at the AN, if } \omega > 90^o.$$

It may be possible to apply a single $\Delta\bar{v}$ to move the apoapsis to coincide with the line-of-nodes prior to the plane change, but the efficiency of that maneuver depends on the values of $\bar{v}$, $\Delta i$, and $\omega$.

If the line-of-nodes and line-of-apsides don't coincide, the velocity vector will have components in both the radial and theta directions at the nodes. To change the inclination, only the theta component needs to be rotated, since the radial component will remain unchanged. This will result in a smaller $\Delta\bar{v}$ than if the entire velocity vector was rotated.

Example 18.1

A satellite is launched from KSC having an inclination of 28.5°, a semi-major axis of 7,400 km, and an eccentricity of 0.05.

(a) Calculate the magnitude of the minimum single Δv required to transfer the satellite to an equatorial orbit having the same size and shape.

(b) If the satellite is equipped with a thruster which provides a fixed Δv of 2.0 $\frac{km}{s}$, find the maximum inclination change possible from its original orbit using a single firing of the thruster.

Solution:

(a) $\Delta i = i_1 - i_2 = 28.5^o - 0.0^o = 28.5^o$

Δv must be performed at apoapsis since the satellite speed is smaller

$$r = a(1 + e) = 7{,}400(1 + 0.05) = 7{,}700 \text{ km}$$

$$v_a = \left[\mu\left(\frac{2}{r} - \frac{1}{a}\right)\right] = \left[3.986 \times 10^5 \left(\frac{2}{7{,}700} - \frac{1}{7{,}400}\right)\right] = 6.98 \ \frac{km}{s}$$

$$\Delta v = 2v_a \sin\left(\frac{\Delta i}{2}\right) = 2(6.98) \sin\left(\frac{28.5^o}{2}\right)$$

$$\Rightarrow \quad \Delta v = 3.44 \ \frac{km}{s}$$

(b) $\Delta i = 2\sin^{-1}\left(\frac{\Delta v}{2v_a}\right) = 2(6.98)\sin^{-1}\left(\frac{2.0}{2(6.98)}\right)$

$$\Rightarrow \quad \Delta i = 16.47^o$$

Example 18.2

A spacecraft is in a circular equatorial Earth orbit at an altitude of 300 km. Using a single impulsive velocity change, the spacecraft is to transfer to a polar orbit having e = 0.02 and a perigee directly above the North Pole. Determine the spacecraft's altitude at perigee and the magnitude of the velocity change required to accomplish this maneuver.

Solution:

$$r = R_E + (alt) = 6.378.1 + 300 = 6{,}678.1 \text{ km}$$

$$v_c = \sqrt{\frac{\mu}{r}} = \sqrt{\frac{3.986\times10^5}{6{,}678.1}} = 7.73\ \frac{\text{km}}{\text{s}}$$

$$a = \frac{r}{1-c^2} = \frac{6{,}678.1}{1-(0.02)^2} = 6{,}680.8 \text{ km}$$

$$r_p = a(1-e) = 6{,}680.8(1-0.02) = 6{,}547.2 \text{ km}$$

$$(alt)_p = r_p - R_E = 6{,}547.2 - 6{,}378.1$$

$$\Longrightarrow\ (alt)_p = 169.1 \text{ km}$$

$$v = \sqrt{\mu\left(\frac{2}{r}-\frac{1}{a}\right)} = \sqrt{(3.986\times10^5)\left(\frac{2}{6{,}678.1}-\frac{1}{6{,}680.8}\right)} = 7.73\ \frac{\text{km}}{\text{s}}$$

Since $v_c$ and v are 90° apart,

$$\Delta v = \sqrt{v_c^2 + v^2} = \sqrt{(7.73)^2 + (7.73)^2}$$

$$\Longrightarrow\ \Delta v = 10.93\ \frac{\text{km}}{\text{s}}$$

<u>Changing the Size and Inclination of an Orbit</u>

Transfer between two <u>non-coplanar</u> circular orbits having different sizes requires both the inclination <u>and</u> size to be changed. This is accomplished by establishing a transfer ellipse and using two velocity changes, one at each end of the ellipse (assuming the line-of-nodes and line-of-apsides coincide).

For maximum efficiency, a portion of the plane change should be performed at each end of the transfer ellipse.

Consider the case of a transfer from orbit 1 to orbit 2, having a different size and inclination

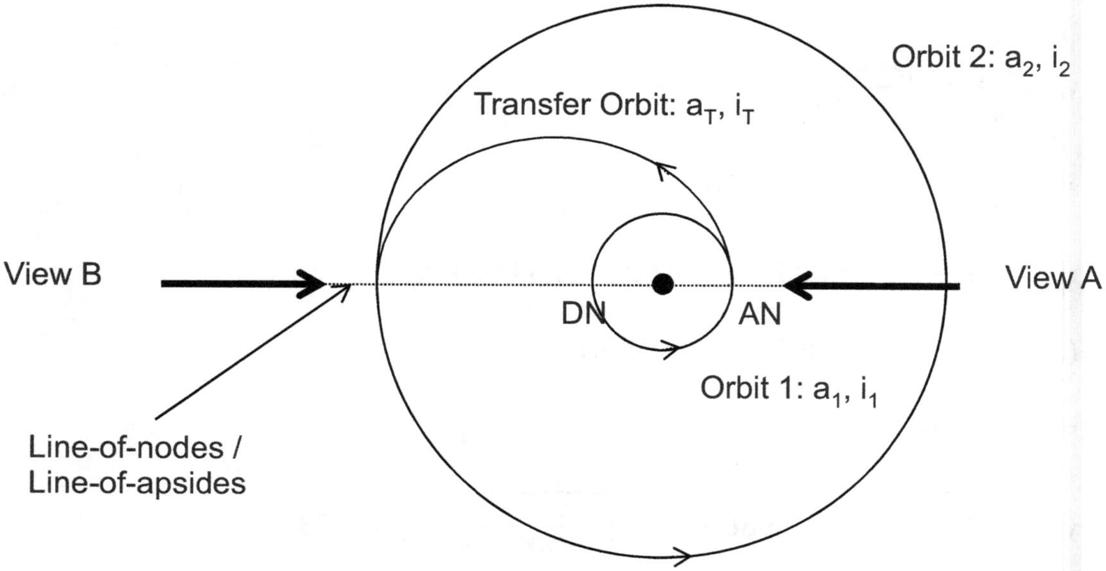


In this case, part of the required plane change should be performed at the periapsis of the transfer orbit during $\Delta\bar{v}_1$, and the remainder of the plane change should be performed at the apoapsis of the transfer orbit during $\Delta\bar{v}_2$. In doing so, the transfer orbit will have an inclination, $i_T$, different from both orbit 1 and orbit 2.

The velocity $\bar{v}_1$ from view A at the AN shows

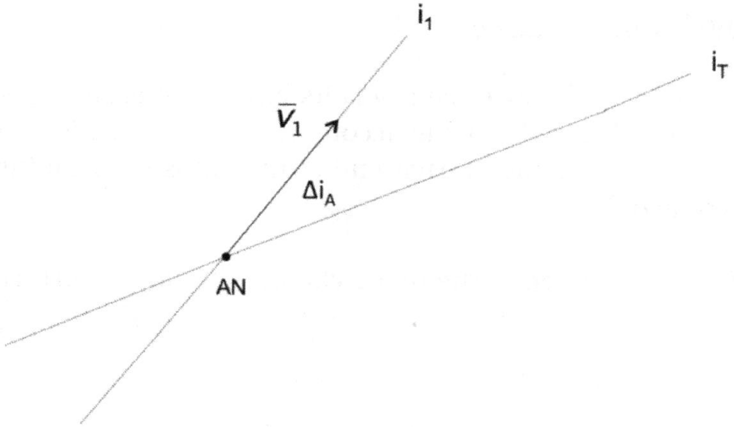

The velocity change, $\Delta\bar{v}_A$, will be applied to both increase the speed and rotate the orbital plane by the inclination $\Delta i_A$, as

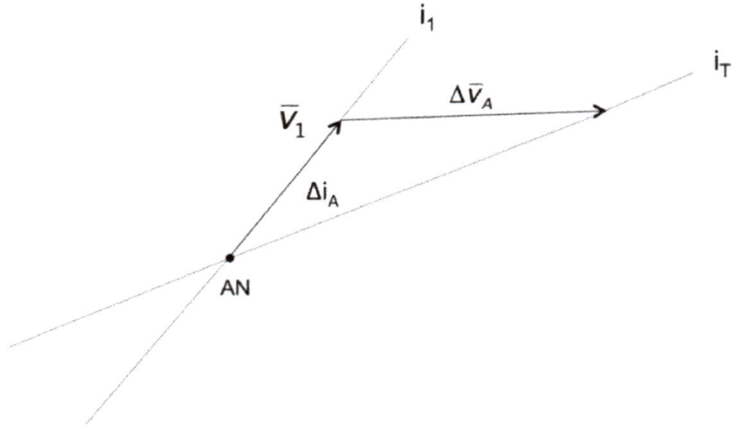


This velocity change yields the result

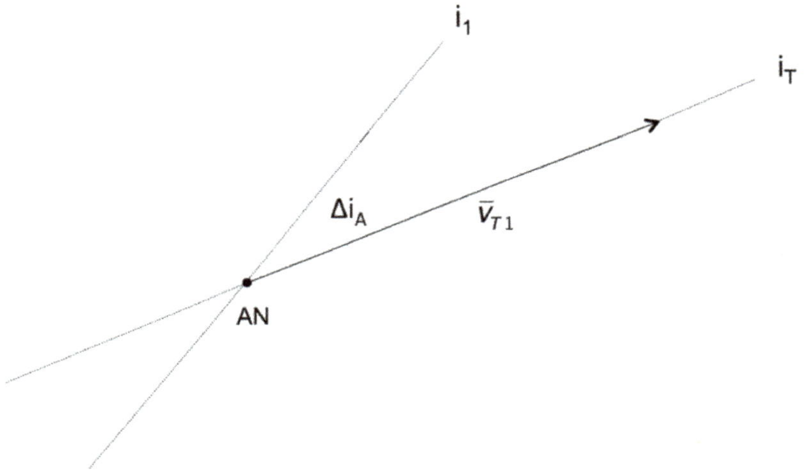

Using the Law of Cosines, the magnitude of $\Delta\bar{v}_A$ can be calculated by

$$v_1 = \sqrt{\frac{\mu}{a_1}}$$

$$a_T = \frac{a_1+a_2}{2}$$

$$v_{T1} = \mu\sqrt{\frac{2}{a_1} + \frac{1}{a_T}}$$

$$\Delta v_A = \sqrt{v_1^2 + v_{T1}^2 - 2v_1 v_{T1}\cos(\Delta i_A)} \qquad \text{(where } \Delta i_A \text{ is unknown)}$$

The rest of the plane change must occur at the periapsis of the transfer ellipse. View B at the DN shows velocity $\bar{v}_{T_2}$

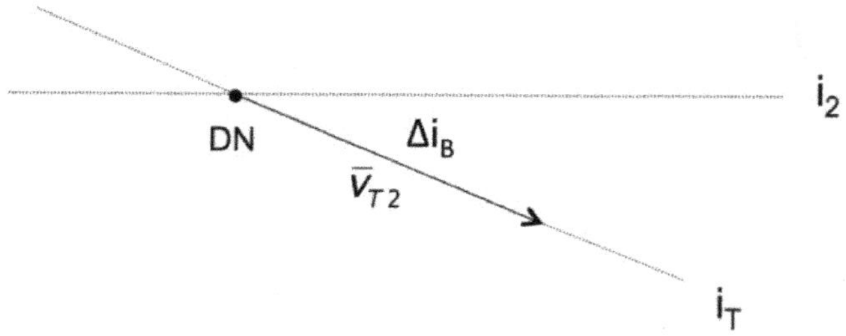


The required velocity change to complete the plane change and the transfer to orbit 2 is

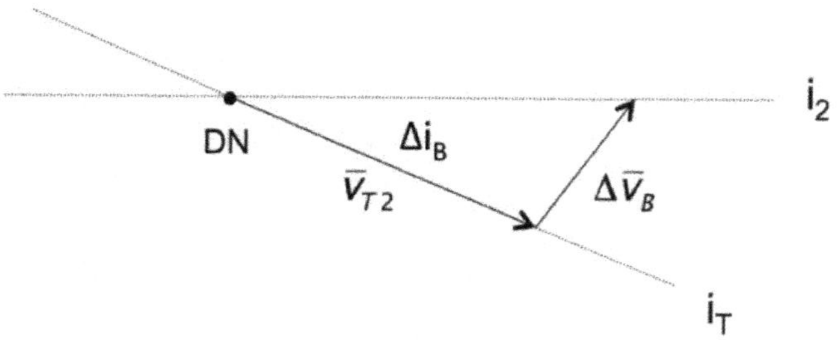

which gives the result

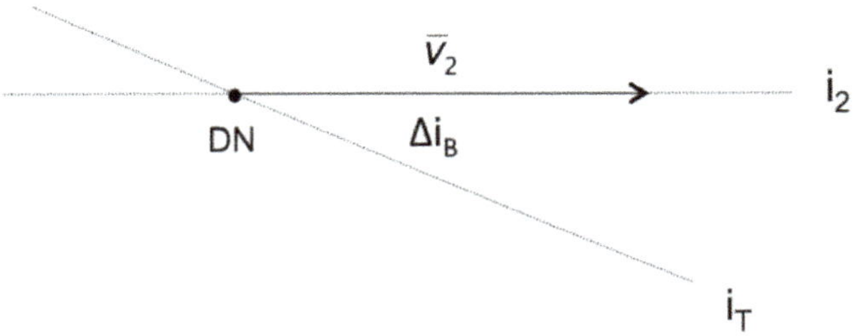


Using the Law of Cosines, $\Delta v_B$ can be calculated as

$$v_2 = \sqrt{\frac{\mu}{a_2}}$$

$$a_T = \frac{a_1 + a_2}{2}$$

$$v_{T2} = \mu\sqrt{\frac{2}{a_2} + \frac{1}{a_T}}$$

$$\Delta v_B = \sqrt{v_2^2 + v_{T2}^2 - 2v_2 v_{T2}\cos(\Delta i_B)} \qquad \text{(where } \Delta i_B \text{ is also unknown)}$$

The magnitude of the total $\Delta v$ for the transfer will be

$$\Delta v = \Delta v_A + \Delta v_B$$

where

$$\Delta i = \Delta i_A + \Delta i_B$$

However, the values of $\Delta i_A$ and $\Delta i_B$ must be determined. To do so, $\Delta i_A$ can be selected to minimize the total $\Delta v$ required for the transfer. This must be performed by the iteration described below

- select several values of $\Delta i_A$ and compute the corresponding value of $\Delta v$
- plot the curve for $\Delta v$ vs , $\Delta i_A$ and select that value of $(\Delta i_A)_{OPT}$ from the curve corresponding to $(\Delta v)_{MIN}$
- once $(\Delta i_A)_{OPT}$ is obtained, calculate $\Delta i_B$ from the expression

$$\Delta i_B = \Delta i - (\Delta i_A)_{OPT}$$

$(\Delta i_A)_{OPT}$ is a function of $a_1$ , $a_2$ , $i_1$ , and $i_2$ , but will generally be a small value on the order of $(\Delta i_A)_{OPT} \sim 3^o$-$5^o$.

The following plot shows an example of the minimization process

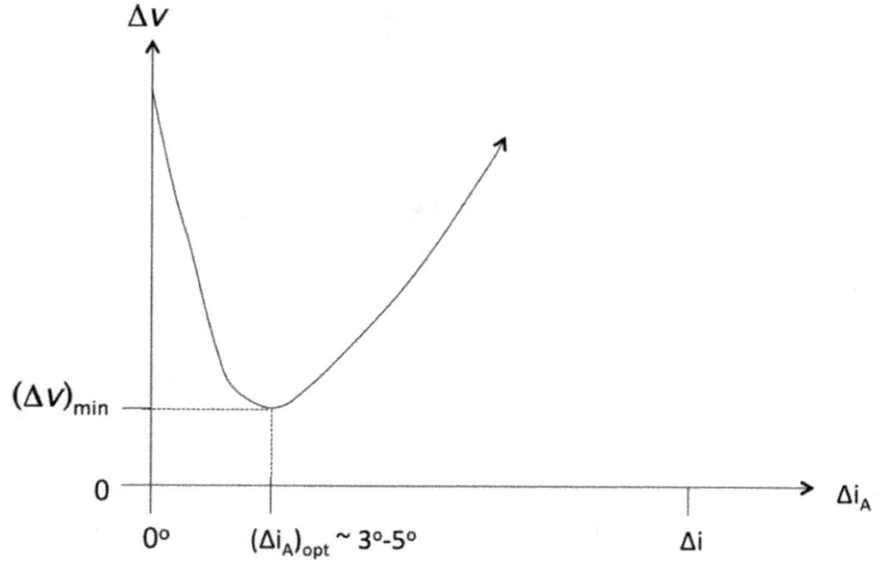


Example 18.3

A communications satellite is launched directly into a 200 km circular Earth orbit having an inclination of 30.06° and must then transfer to a circular, equatorial orbit having an altitude of 6,800 km. Determine the minimum velocity change required to complete the maneuver using a Hohmann-type transfer. Specify the portion of the plane change to be done at both ends of the transfer ellipse for maximum efficiency.

Solution:

$$a_1 = R_E + (alt)_1 = 6{,}378.1 + 200 = 6{,}578.1 \text{ km}$$

$$a_2 = R_E + (alt)_2 = 6{,}378.1 + 6{,}800 = 13{,}178.1 \text{ km}$$

$$a_T = \frac{a_1+a_2}{2} = \frac{6{,}578.1+13{,}178.2}{2} = 9.878.1 \text{ km}$$

$$\Delta i = |i_2 - i_1| = |0.0^o - 30.06^o| = 30.06^o$$

$$V_1 = \sqrt{\frac{\mu}{a_1}} = \sqrt{\frac{3.986\times10^5}{6{,}578.1}} = 7.78 \frac{\text{km}}{\text{s}}$$

$$v_{T1} = \sqrt{\mu\left(\frac{2}{a_1} - \frac{1}{a_T}\right)} = \sqrt{(3.986 \times 10^5)\left(\frac{2}{6{,}578.1} - \frac{1}{9{,}878.1}\right)} = 8.99\ \frac{km}{s}$$

$$v_{T2} = \sqrt{\mu\left(\frac{2}{a_2} - \frac{1}{a_T}\right)} = \sqrt{(3.986 \times 10^5)\left(\frac{2}{13{,}178.1} - \frac{1}{9{,}878.1}\right)} = 4.49\ \frac{km}{s}$$

$$v_2 = \sqrt{\frac{\mu}{a_2}} = \sqrt{\frac{3.986 \times 10^5}{13{,}178.1}} = 5.50\ \frac{km}{s}$$

$$\Delta v_1 = \sqrt{v_1^2 + v_{T1}^2 - 2v_1 v_{T1} \cos \Delta i_p}$$

$$\Delta i = \Delta i_p + \Delta i_a$$

$$\Delta v_1 = \sqrt{v_2^2 + v_{T2}^2 - 2v_2 v_{T2} \cos \Delta i_a}$$

Using the above equations, the following table was created.

| $\Delta i_p$ | $\Delta i_a$ | $\Delta v_1$ | $\Delta v_2$ | $\Delta v$ |
|---|---|---|---|---|
| 0.0° | 30.06° | 1.210 | 2.768 | 3.978 |
| 2.0° | 28.06° | 1.245 | 2.613 | 3.878 |
| 4.0° | 26.06° | 1.343 | 2.458 | 3.801 |
| 6.0° | 24.06° | 1.493 | 2.305 | 3.798 |
| 8.0° | 22.06° | 1.681 | 2.153 | 3.834 |
| 10.0° | 20.06° | 1.895 | 2.004 | 3.899 |

Plotting the results gives the curve below.

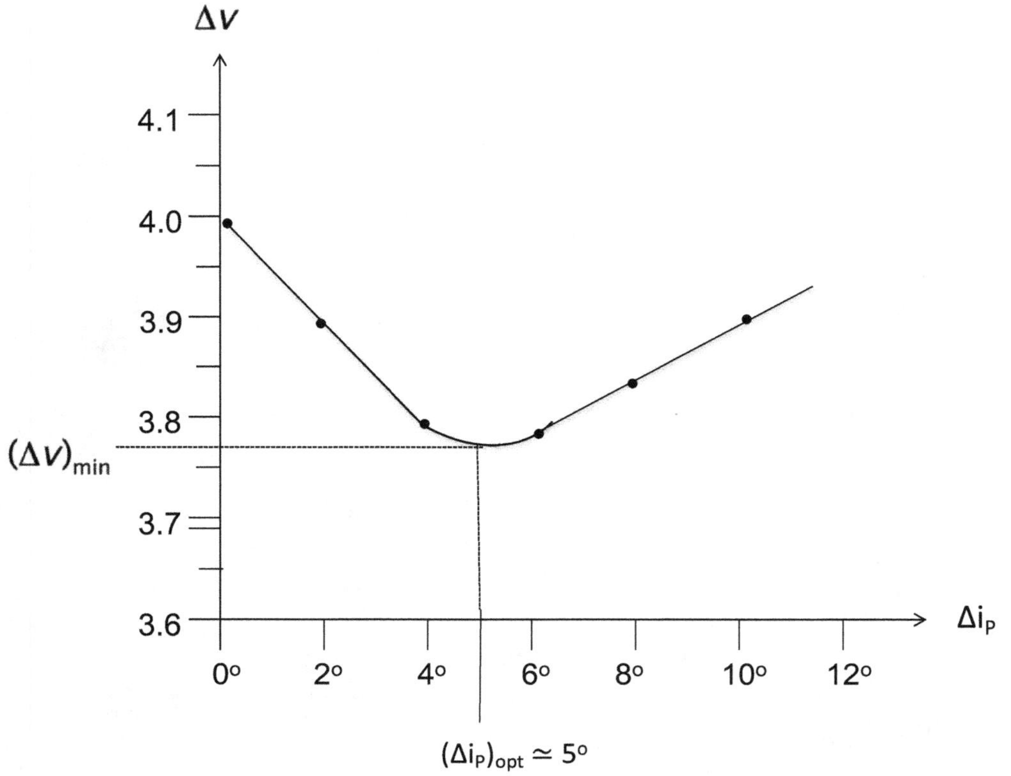


The curve indicates that the minimum $\Delta v$ occurs at a value of $\Delta i_p \approx 5.0^o$. Therefore, the optimal solution becomes

$(\Delta i_p)_{OPT} \approx 5.0^o$ $\qquad$ $(\Delta i_a)_{OPT} \approx 25.06^o$

$\Delta v_1 = 1.413 \frac{km}{s}$ $\qquad$ $\Delta v_2 = 2.381 \frac{km}{s}$

$$\Delta v = \Delta v_1 + \Delta v_2 = 3.794 \frac{km}{s}$$

Problems

18.1 A satellite in an Earth orbit has the following orbital elements

$$a = 9{,}500 \text{ km} \qquad e = 0.15 \qquad h = 60{,}840 \frac{\text{km}^3}{\text{s}^2} \qquad i = 10^\circ \qquad \omega = 45^\circ$$

Compute the minimum single velocity change required to increase the inclination to 30°.
(Ans. $\Delta v = 1.02 \frac{\text{km}}{\text{s}}$)

18.2 An Earth satellite is launched into an orbit having an altitude of 732 km at periapsis and an eccentricity of 0.10. The satellite is equipped with a fixed $\Delta v$ thruster capable of supplying an instantaneous velocity change of 2.4 $\frac{\text{km}}{\text{s}}$. Determine the maximum increase in orbital inclination that can be achieved with a single burn of the thruster. Assume the line-of-nodes and line-of-apsides coincide.
(Ans. $\Delta i = 21.53^\circ$)

18.3 Consider that a satellite is to transfer from a 500 km altitude circular equatorial orbit to a 300 km altitude circular orbit having an inclination of 10° using a Hohmann transfer. If all the plane change is to occur at the apogee of the transfer ellipse, calculate the magnitude of the first $\Delta v$ required in this transfer. Assume the line-of-nodes and line-of-apsides coincide.
(Ans. $\Delta v = 1.32 \frac{\text{km}}{\text{s}}$)

18.4 A satellite in an Earth orbit has the following orbital elements

$$a = 10{,}00 \text{ km} \qquad e = 0.1 \qquad i = 10^\circ \qquad \omega = 30^\circ$$

Compute the minimum single velocity change required to increase the inclination to 25°.
(Ans. $\Delta v = 1.51 \frac{\text{km}}{\text{s}}$)

18.5 Consider an Earth satellite having the following orbital elements

$$a = 15{,}00 \text{ km} \qquad e = 0.5 \qquad i = 10^\circ \qquad \Omega = 45^\circ \qquad \omega = 30^\circ$$

Find the minimum single velocity change required to reduce the inclination to 0°.
(Ans. $\Delta v = 1.49 \frac{\text{km}}{\text{s}}$)

18.6 Using a single impulsive $\Delta v$, an Earth satellite in a 400 km circular orbit is to change from an inclination of $60^{o}$ to an elliptical orbit having an inclination of $40^{o}$ and an eccentricity of 0.5. Calculate the minimum $\Delta v$ to accomplish this maneuver.
(Ans. $\Delta v = 3.41 \frac{km}{s}$)

18.7 A satellite is to transfer from a circular Earth orbit having $a = 7{,}000$ km and $i = 30^{o}$, to a circular equatorial orbit having $a = 12{,}000$ km. Consider that $5^{o}$ of the plane change is to be done at the periapsis of the transfer orbit and the remainder to be done at the apoapsis. Compute the magnitude of the total velocity change required to make this transfer.
(Ans. $\Delta v = 3.62 \frac{km}{s}$)

18.8 An Earth satellite is in a circular orbit at an altitude of 200 km and an inclination of $28.5^{o}$. Determine the <u>minimum</u> $\Delta v$ required to transfer this satellite to a circular, geostationary orbit. Consider that the line-of-nodes and the line-of-apsides coincide and determine the optimal amount of the plane change to be performed at both the periapsis and apoapsis of the transfer orbit.
(Ans. $(\Delta v)_{min} = 4.27 \frac{km}{s}$, $(\Delta i_L)_{OPT} \approx 2.3^{o}$, $(\Delta i_H)_{OPT} \approx 26.2^{o}$)

# Module 19: Phasing Maneuvers

A phasing maneuver is a two impulse Hohmann transfer from and back to the same orbit, which can move a spacecraft forward or backward within its orbit.

Phasing maneuvers are used for rendezvous with another spacecraft in the same orbit, and for station-keeping maneuvers.

A phasing orbit is selected to have a period, which will return the spacecraft to its original orbit within a specified time.

## Moving Backward in an Orbit

Consider the figure below where a spacecraft is originally in position A of orbit 1

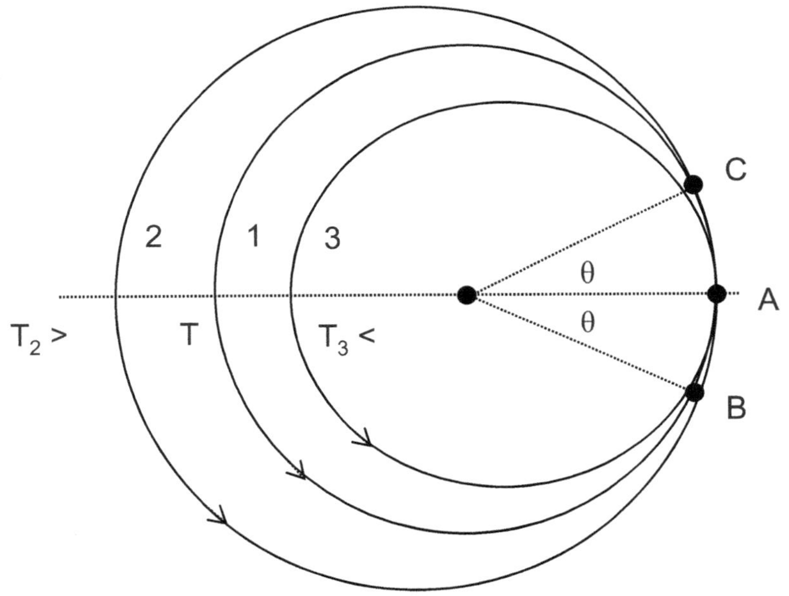

Consider moving a spacecraft backward from point A to point B in a circular orbit.

- Apply a tangential $\Delta\bar{v}_1$ at point A in the direction of travel to move the spacecraft into an elliptical orbit having a period, $T_2 > T_1$.

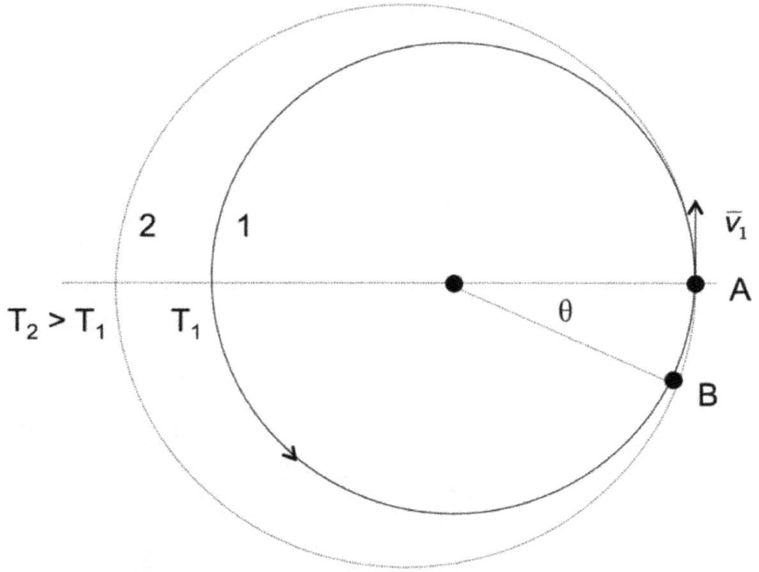


- This will increase the velocity of the spacecraft

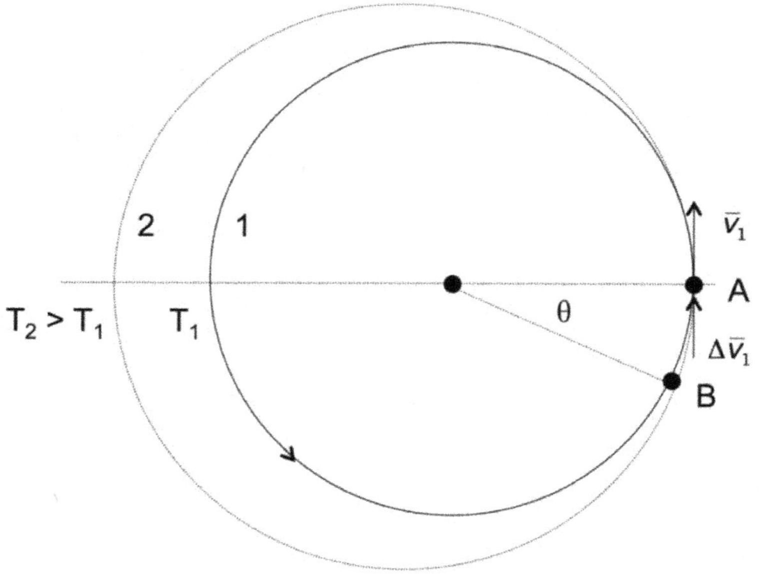

- Increasing the velocity to $\bar{v}_2$ will move the spacecraft to orbit 2 having a longer period, $T_2$

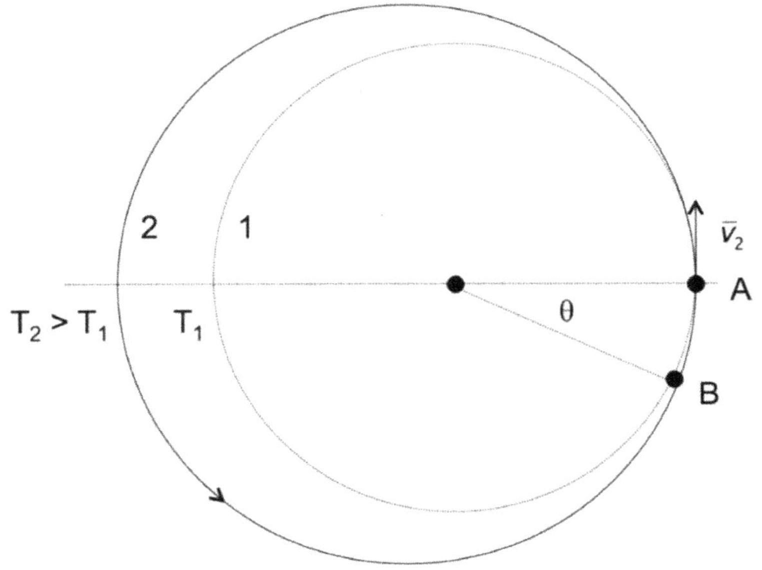


The spacecraft will make one complete revolution in orbit 2 and return to original position with velocity $\bar{v}_2$

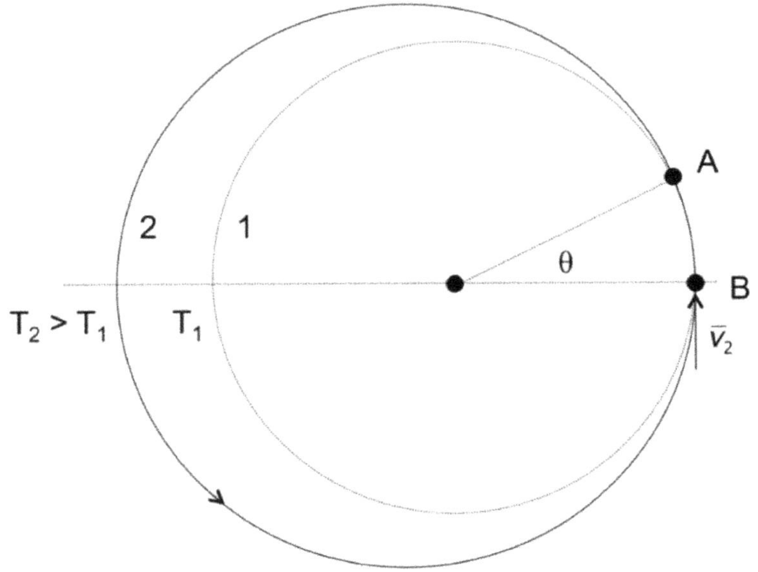

- Then, apply a tangential $\Delta\bar{v}_2$ in a direction opposite to $\Delta\bar{v}_2$ at the original position to decrease speed to re-establish the original circular orbit 1

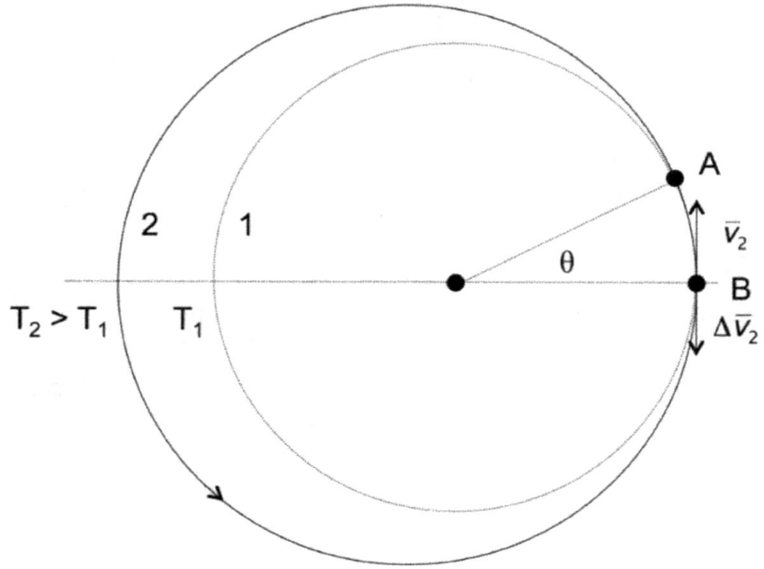


- The final position of the spacecraft will coincide with the original position B having its original velocity, $\bar{v}_1$, but in a position an angle θ behind the original position A

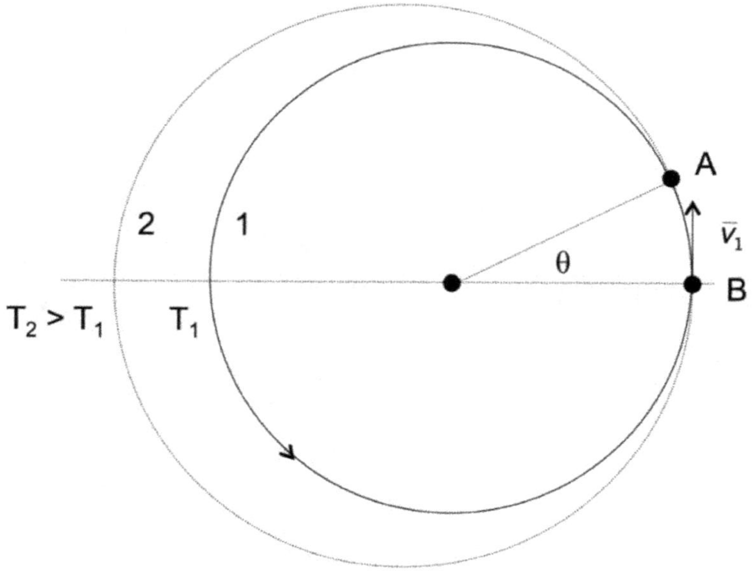


The overall effect of the maneuver is that the spacecraft increases its velocity to move backward in its orbit.

Example 19.1

A spacecraft in a circular Earth orbit at radius r launches a space probe to study the Van Allen radiation belts. The probe will complete one elliptical orbit and return to the point where it was launched and rendezvous with the spacecraft after the spacecraft completes 100 revolutions in its original orbit. In terms of r, determine
(a) the semi-major axis of the elliptical orbit
(b) the total velocity change applied to the probe to complete this mission.

Solution:

(a) $T_{SC} = 2\pi\sqrt{\frac{r^3}{\mu}}$

$$T_P = 2\pi\sqrt{\frac{a^3}{\mu}}$$

$$T_P = 100T_{SC}$$

$$2\pi\sqrt{\frac{a^3}{\mu}} = 100(2\pi)\sqrt{\frac{r^3}{\mu}}$$

$$a = \sqrt[3]{(100)^2}\ r$$

$$\Rightarrow\ a = 21.54\ r$$

(b) $v_{SC} = \sqrt{\frac{\mu}{r}}$

$$v_P = \sqrt{\mu\left(\frac{2}{r}-\frac{1}{a}\right)} = \sqrt{\mu\left(\frac{2}{r}-\frac{1}{21.54r}\right)}\sqrt{\mu\left(\frac{2(21.54)}{21.54r}-\frac{1}{21.54r}\right)} = 1.398\sqrt{\frac{\mu}{r}}$$

$$\Delta v = 2(v_P - v_{SC}) = 2\left(1.398\sqrt{\frac{\mu}{r}} - \sqrt{\frac{\mu}{r}}\right)$$

$$\Rightarrow\ \Delta v = 0.80\sqrt{\frac{\mu}{r}}$$

Example 19.2

For the problem described in Example 19.1, derive an expression for $\Delta v$ (in terms of $v_{SC}$) required for the probe as a function of the number of revolutions, n, that the spacecraft completes in its original orbit. Evaluate the expression for the cases where n = 100, 10, 5, and 2.

Solution:

$$T_{SC} = 2\pi\sqrt{\frac{r^3}{\mu}}$$

$$T_P = 2\pi\sqrt{\frac{a^3}{\mu}}$$

$$T_P = nT_{SC}$$

$$2\pi\sqrt{\frac{a^3}{\mu}} = n(2\pi)\sqrt{\frac{r^3}{\mu}}$$

$$a = \sqrt[3]{n^2}\, r = n^{2/3}r$$

$$v_P = \sqrt{\mu\left(\frac{2}{r}-\frac{1}{a}\right)} = \sqrt{\mu\left(\frac{2}{r}-\frac{1}{n^{2/3}r}\right)}\sqrt{\mu\left(\frac{2}{r}-\frac{1}{n^{2/3}r}\right)} = \sqrt{2-\frac{1}{n^{2/3}}}\sqrt{\frac{\mu}{r}}$$

$$v_{SC} = \sqrt{\frac{\mu}{r}}$$

$$\Delta v = 2(v_P - v_{SC}) = 2\left[\sqrt{2-\frac{1}{n^{2/3}}}\sqrt{\frac{\mu}{r}} - \sqrt{\frac{\mu}{r}}\right] = 2\left[\sqrt{\frac{2n^{2/3}-1}{n^{2/3}}} - 1\right]\sqrt{\frac{\mu}{r}}$$

$$\Rightarrow \Delta v = 2\left[\sqrt{\frac{2n^{2/3}-1}{n^{2/3}}} - 1\right]v_{SC}$$

$n = 100$: $\Delta v = 0.796\ v_{SC}$

$n = 10$: $\Delta v = 0.672\ v_{SC}$

$n = 5$: $\Delta v = 0.575\ v_{SC}$

$m = 2$: $\Delta v = 0.341\ v_{SC}$

Moving Forward in an Orbit

Consider moving a spacecraft forward from point A to point C in a circular orbit

- Apply a tangential velocity change opposite the direction of travel to move the spacecraft into an elliptical orbit having, $T_3 < T_1$.

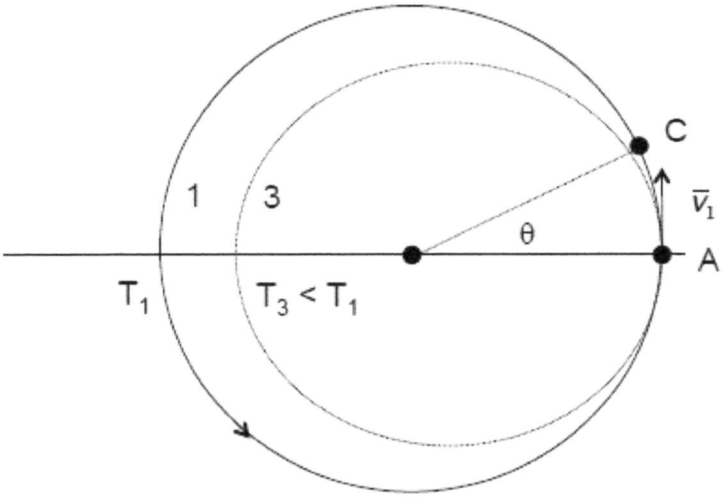


- This will decrease the speed of the spacecraft

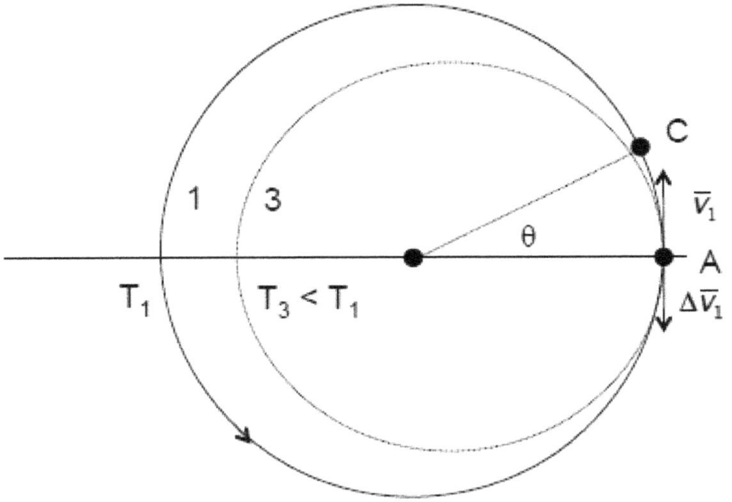

- Decreasing the velocity to $\bar{v}_3$ will move the spacecraft to orbit 3 having a shorter period, $T_3$

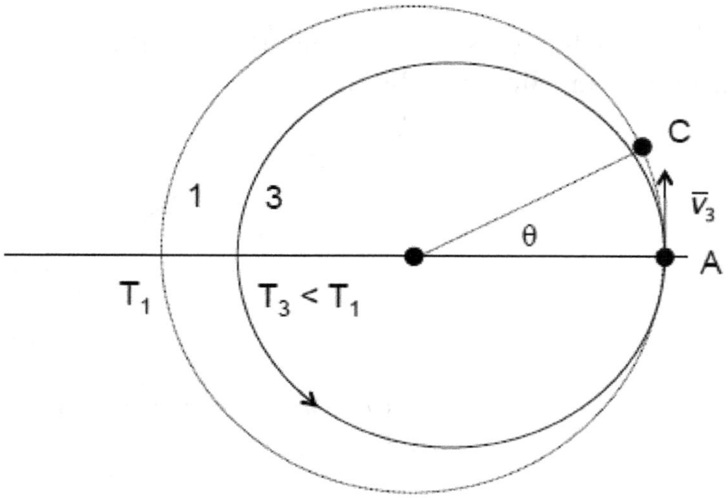


- The spacecraft will make one complete revolution in orbit 3 and return to original position

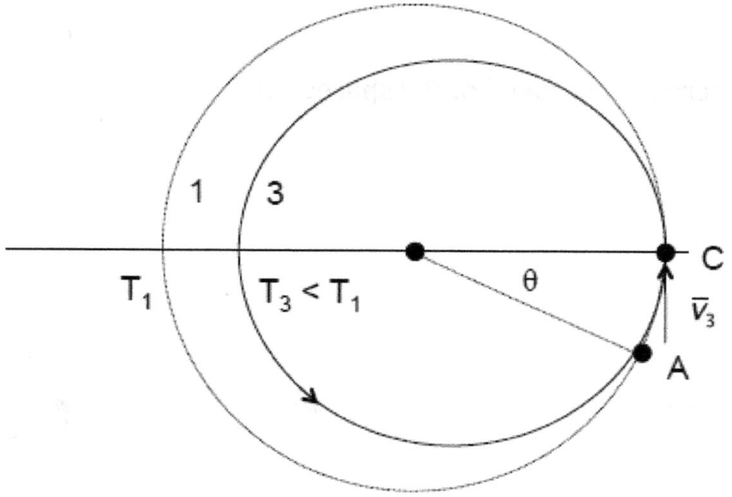

- Apply a tangential $\Delta\bar{v}_3$ at original position to increase speed to re-establish original circular orbit 1

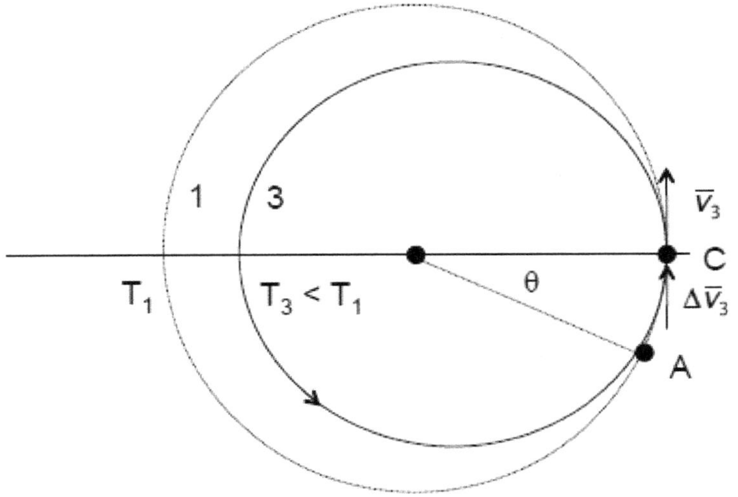


- The final position of the spacecraft will coincide with the original position C, having its original velocity, $\bar{v}_1$, but in a position an angle θ ahead of the original position A

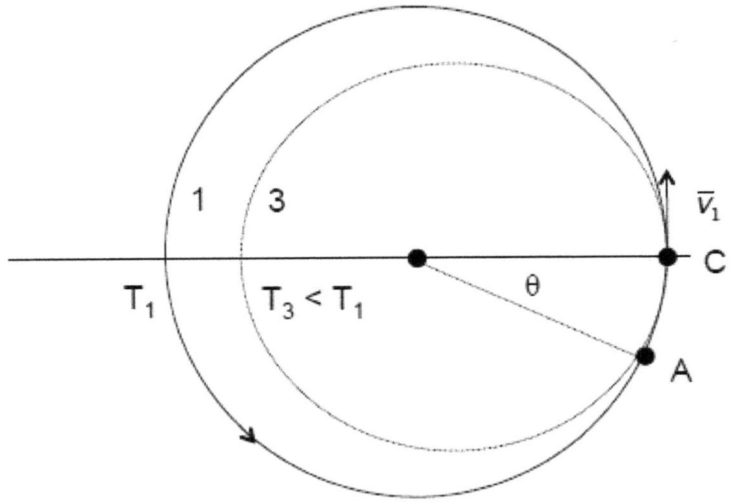


The overall effect of the maneuver is that the spacecraft decreases its velocity to move forward in its orbit.

Summary of Phasing Maneuvers

- Determine the amount of distance or time that is desired to move forward or backward in orbit.

- Calculate the period of the appropriate phasing orbit.

- Compute the size of the phasing orbit.

- Calculate the $\Delta\bar{v}$ required to move onto the phasing orbit then back to the original orbit.

Example 19.3

Earth satellites A and B are in the same circular orbit of radius r but are located 180° apart. A single Δv is to be applied to satellite A to perform a phasing maneuver so both A and B will arrive back at the original position of satellite A at the same time. The phasing orbit is to be designed so that A will travel 4 complete revolutions in its phasing orbit while B travels 2½ complete revolutions in its original orbit before meeting at the desired location. Determine
(a) the semi-major axis of the phasing orbit in terms of r of the original orbit
(b) the Δv required to achieve this maneuver in terms of the original circular orbital speed.

Solution:

$$\text{(a)}\ T_A = 2\pi\sqrt{\frac{a^3}{\mu}}$$

$$T_B = 2\pi\sqrt{\frac{r^3}{\mu}}$$

$$4T_A = 2.5T_B$$

$$8\pi\sqrt{\frac{a^3}{\mu}} = 5\pi\sqrt{\frac{r^3}{\mu}}$$

$$\Rightarrow\ a = 0.73r$$

(b) $v_c = \sqrt{\frac{\mu}{r}}$

$$v_{T_1} = \sqrt{\mu\left(\frac{2}{r} - \frac{1}{a}\right)} = \sqrt{\mu\left(\frac{2}{r} - \frac{1}{0.73r}\right)} = \sqrt{\frac{0.63\mu}{r}} = 0.79\sqrt{\frac{\mu}{r}} = 0.79v_c$$

$$\Delta v = v_{T_1} - v_c = 0.79v_c - v_c$$

$$\Longrightarrow \quad \Delta v = -0.21v_c$$

Problems

19.1 Satellites A and B are in the same circular orbit having a radius of r. In this orbit, B is located 180° ahead of A.
(a) Calculate the semi-major axis (in terms of r) of a phasing orbit in which satellite A will re-position itself to a location 30° behind of satellite B in the same original orbit after A travels 5 complete orbits and B travels at least 2 complete orbits.
(b) Find the total Δv required (in terms of the circular orbit speed of B) to compete this phasing maneuver for re-positioning A to the specified location.
(Ans. (a) $a = 0.64r$; (b) $\Delta v = 0.66v_B$)

19.2 Satellites A and B are in the same circular orbit having a radius of r. In this orbit, A is located 180° behind B. Determine the semi-major axis (in terms of r) of a phasing orbit in which satellite A will rendezvous with B after just one revolution in the phasing orbit.
(Ans. $a = 0.63r$)

19.3 Satellites A and B are in the same circular orbit of radius r, with B positioned 90° behind A. Compute the semi-major axis of the phasing orbit (in terms of r) in which A will rendezvous with B after just one complete revolutions in the phasing orbit. Rendezvous will occur at the original position of A, while B travels through 2¼ revolutions in its original orbit.
(Ans. $a = 1.72r$)

19.4 Calculate the Δv requirement to shift a geostationary orbit 9° westward in 3 revolutions of its phasing orbit.
(Ans. $\Delta v = 0.16\ \frac{km}{s}$)

19.5 Find the Δv requirement to shift a geostationary orbit 9° eastward in 3 revolutions of its phasing orbit.
(Ans. $\Delta v = 0.02\ \frac{km}{s}$)

19.6 Consider an elliptical orbit where $r_p = 7{,}000$ km, $r_a = 14{,}000$ km. Satellite A is positioned at perigee and Satellite B is 90° ahead of A. Calculate the $\Delta v$ required for chaser A to catch B at A's original location after one revolution in a phasing orbit.
(Ans. $\Delta v = 0.48\ \frac{km}{s}$)

19.7 Satellites A and B are in the same circular orbit of radius r and B is located 180° ahead of A. Determine
(a) the semi-major axis (in terms of r) of a phasing orbit in which satellite A will re-position itself to a new location 60° ahead of satellite B in the same original orbit after A travels 6 complete orbits and B travels at least 3 complete orbits
(b) the total $\Delta v$ required (in terms of the circular orbit speed of satellite B) to complete the phasing maneuver for re-positioning satellite A to the specified location.
(Ans. (a) $a = 0.68r$; (b) $\Delta v = 0.56v_B$)

19.8 Two spacecraft, A and B, are in the same circular Earth orbit at an altitude of 350 km. The vehicles are separated by 1,200 km with spacecraft B leading. Both spacecraft apply a tangential $\Delta v$ at the same instant so that they can rendezvous at a point midway between their initial positions in one revolution of their phasing orbits. Compute
(a) the times required for each spacecraft to reach the rendezvous point
(b) the total $\Delta v$ requirement for each spacecraft.
(Ans. (a) $t_A = 90.2$ min, $t_B = 92.8$ min; (b) $\Delta v_A = 0.04\ \frac{km}{s}$, $\Delta v_B = 0.04\ \frac{km}{s}$)

# Module 20: Interplanetary Trajectories - The Patched Conic Method

There are many important aspects to consider in the accurate design and analysis of interplanetary trajectories. The first is the definition of the sphere of influence (SOI) of a planet, which is the radius from the center of the planet beyond which the planet's gravitational influence is less than that of the Sun.

## Sphere of Influence

The SOI is defined as the sphere around a planet beyond which the gravitational attraction of the planet is considered to be negligible, i.e., the point considered to be 'infinity' for the purpose of planetary departure. The SOI is calculated using the formula

$$r_{SOI} = a_{PL}\left[\frac{m_{PL}}{m_{Sun}}\right]^{2/5}$$

where

$a_{PL}$ = semi-major axis of the planet's orbit about the Sun
$m_{PL}$ = mass of the departure/arrival planet
$m_{Sun}$ = mass of the Sun

The spheres of influence of the planets are

| Planet | $r_{SOI}$ (km) |
|---|---|
| Mercury | $1.12 \times 10^5$ |
| Venus | $6.16 \times 10^5$ |
| Earth | $9.25 \times 10^5$ |
| (Moon) | $6.61 \times 10^4$ |
| Mars | $5.77 \times 10^5$ |
| Jupiter | $4.82 \times 10^7$ |
| Saturn | $5.48 \times 10^7$ |
| Uranus | $5.18 \times 10^7$ |
| Neptune | $8.66 \times 10^7$ |
| Pluto | $3.08 \times 10^6$ |

Example 20.1

Calculate the radii of the spheres of influence of Uranus and Pluto.

Solution:

$$r_{SOI_{Uranus}} = a_U \left[\frac{m_U}{m_S}\right]^{2/5} = 2.872 \times 10^9 \left[\frac{8.863\times10^{25}}{1.989\times10^{30}}\right]^{2/5} = 5.18 \times 10^7 \text{ km}$$

$$r_{SOI_{Pluto}} = a_P \left[\frac{m_P}{m_S}\right]^{2/5} = 5.870 \times 10^9 \left[\frac{1.250\times10^{22}}{1.989\times10^{30}}\right]^{2/5} = 3.08 \times 10^6 \text{ km}$$

Patched Conic Method

Interplanetary travel is modeled as a series of two-body problems, patched together to form a complete trajectory. Each two-body problem consists of the spacecraft and whichever body's gravity has the greatest influence at the time, where all bodies are considered to be point masses.

The Patched Conic Method for interplanetary trajectories can be described in three phases:

1. Planet departure: Establishes the required $\Delta v$ to escape the departure planet's SOI on a hyperbolic trajectory. (This phase will be described in Modules 21 and 22 for the possible transfer scenarios.) When the departure planet's SOI is reached, the departure planet's gravity is turned-off and the Sun's gravity is turned-on.

2. Heliocentric transfer between planets: The spacecraft moves on a trajectory from the departure planet to the target planet, often using a Hohmann transfer for fuel efficiency. During interplanetary travel a spacecraft spends most of its time under the influence of the Sun.

3. Planet arrival: When the target planet's SOI is reached, the Sun's gravity is turned-off, the arrival planet's gravity is turned-on, and the spacecraft approaches the arrival planet on a hyperbolic trajectory. (This phase will be described in Modules 21-23 for the possible transfer scenarios.)

In short, this method ignores the gravitational attraction of all bodies except the one whose influence is the greatest. This approximation is extremely accurate for preliminary mission design and mission planning activities.

The Patched Conic Method is used for travel to both superior and inferior planets. A superior planet is defined as a planet, which is farther away from the Sun than Earth, while an inferior planet is defined as a planet, which is closer to the Sun than Earth.

A typical patched conic trajectory for a transfer to a superior planet is shown in the following figure

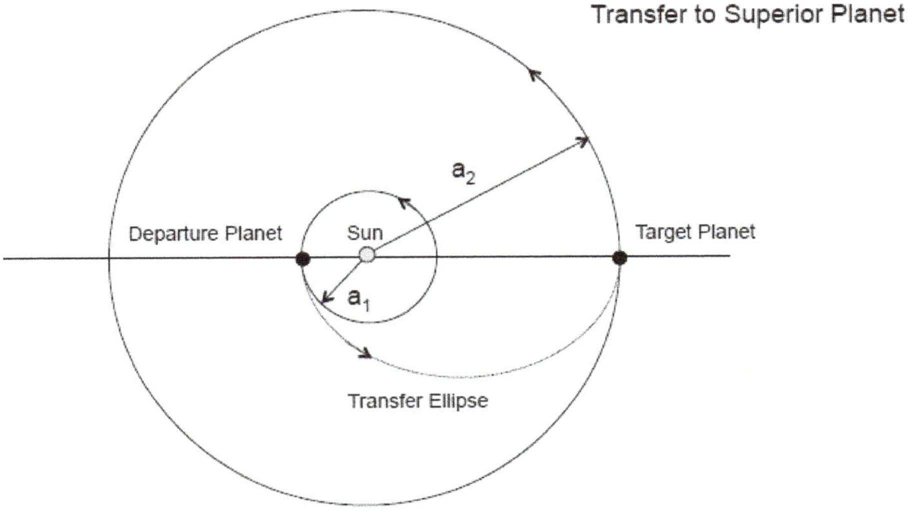


- The spacecraft needs to speed-up relative to the Sun, so it must depart in the direction of the velocity of the planet relative to the Sun. It must also speed-up relative to the planet to escape the planet's gravity.

A typical patched conic trajectory for a transfer to an inferior planet is shown in the following figure

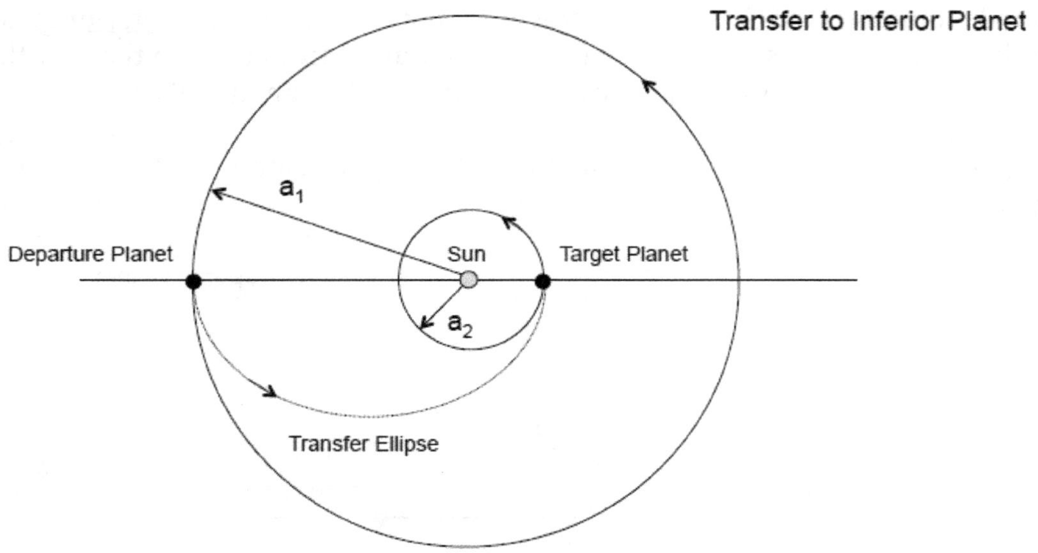


- The spacecraft needs to slow-down relative to the Sun, so it must depart in the direction opposite to the velocity of the planet relative to the Sun. It must still speed-up relative to the planet to escape the planet's gravity.

It should be noted that for both transfers, the departure direction on the transfer ellipse (relative to the Sun) is in the same direction as the planet orbits about the Sun.

Synodic Period

Timing is the most important consideration for interplanetary travel. The phase angle at departure defines the proper position of the planets to achieve an interplanetary trajectory.

The 'synodic period', $\tau_S$, is defined as the time for the required phase angle to repeat itself, which indicates that a launch can occur once every $\tau_S$ for a specified trajectory.

The synodic period is calculated from the expression

$$\tau_S = \frac{2\pi}{|\omega_D - \omega_T|}$$

where

$\omega_D$ = magnitude of the angular velocity of the departure planet
$\omega_T$ = magnitude of the angular velocity of the target planet

If a departure time is missed, the mission must wait another $\tau_S$ or compute a new trajectory for an alternate departure time.

Examples of synodic periods of planets with Earth show

| Planet | $\tau_S$ (yr) |
|---|---|
| Mercury | 0.32 |
| Venus | 1.60 |
| Mars | 2.13 |
| all others | ~ 1.0 |

Example 20.2

For an interplanetary Hohmann transfer from Neptune to Mercury and assuming circular orbits for both planets, calculate the following quantities
(a) the synodic period in Earth days, and
(b) the required phase angle at departure.

Solution:

(a) $$T_N = 2\pi\sqrt{\frac{a_N^3}{\mu_{Sun}}} = 2\pi\sqrt{\frac{(4.495\times10^9)^3}{1.327\times10^{11}}} = 5.198 \times 10^9 \text{ sec}$$

$$T_M = 2\pi\sqrt{\frac{a_M^3}{\mu_{Sun}}} = 2\pi\sqrt{\frac{(5.791\times10^7)^3}{1.327\times10^{11}}} = 7.601 \times 10^6 \text{ sec}$$

$$\tau_S = \frac{T_N T_M}{|T_N - T_M|} = \frac{(5.198\times10^9)(7.601\times10^6)}{|5.198\times10^9 - 7.601\times10^6|}$$

$$\Rightarrow \quad \tau_S = 7.612 \times 10^6 \text{sec} = 0.241 \text{ yr} = 88.1 \text{ days}$$

(b) $$a_T = \frac{(a_N + a_M)}{2} = \frac{(4.495\times10^9 + 5.791\times10^7)}{2} = 2.2765 \times 10^9 \text{ km}$$

$$t_T = \frac{T_T}{2} = \frac{2\pi}{2}\sqrt{\frac{a_T^3}{\mu_{Sun}}} = \pi\sqrt{\frac{(2.2765\times10^9)^3}{1.327\times10^{11}}} = 936{,}705{,}383.0 \text{ sec} = 10{,}841.5 \text{ days}$$

$$T_M = 2\pi\sqrt{\frac{a_M^3}{\mu_{Sun}}} = 2\pi\sqrt{\frac{(5.791\times10^7)^3}{1.327\times10^{11}}} = 7{,}601.070.4 \text{ sec} = 87.98 \text{ days}$$

$$\gamma = \frac{t_T}{T_M}(360^o) = \frac{10{,}841.5}{87.795}(360^o) = 44{,}364.18^o$$

$$\Rightarrow \quad \gamma = 84.18^o \quad (\leq 360^o)$$

Example 20.3

Consider an interplanetary Hohmann transfer from Mercury to Pluto. Assuming circular orbits for both planets about the Sun, determine the hyperbolic excess speeds at Mercury departure and at Pluto arrival.

Solution:

$$v_{M/Sun} = \sqrt{\frac{\mu_{Sun}}{a_M}} = \sqrt{\frac{1.327\times10^{11}}{5.791\times10^{7}}} = 47.87\ \frac{km}{s}$$

$$v_{P/Sun} = \sqrt{\frac{\mu_{Sun}}{a_P}} = \sqrt{\frac{1.327\times10^{11}}{5.870\times10^{9}}} = 4.75\ \frac{km}{s}$$

$$a_T = \frac{a_M+a_P}{2} = \frac{5.791\times10^{7}+5.870\times10^{9}}{2} = 2.964\times10^{9}\ km$$

$$v_{T_1} = \sqrt{\mu_{Sun}\left(\frac{2}{a_M}-\frac{1}{a_T}\right)} = \sqrt{(1.327\times10^{11})\left(\frac{2}{5.791\times10^{7}}-\frac{1}{2.964\times10^{9}}\right)} = 67.37\ \frac{km}{s}$$

$$v_{T_2} = \sqrt{\mu_{Sun}\left(\frac{2}{a_P}-\frac{1}{a_T}\right)} = \sqrt{(1.327\times10^{11})\left(\frac{2}{5.870\times10^{9}}-\frac{1}{2.964\times10^{9}}\right)} = 0.67\ \frac{km}{s}$$

$$v^{+}_{\infty/M} = v_{T_1} - v_{M/Sun} = 67.37 - 47.87$$

$$\Rightarrow v^{+}_{\infty/M} = 19.50\ \frac{km}{s}$$

$$v^{-}_{\infty/P} = v_{P/Sun} - v_{T_2} = 4.75 - 0.67$$

$$\Rightarrow v^{-}_{\infty/P} = 4.08\ \frac{km}{s}$$

Problems

20.1 Calculate the radii of the spheres of influence of Mercury and Mars. ($r_{SOI_{Mercury}} = 1.12\times10^{5}$ km, $r_{SOI_{Mars}} = 5.77\times10^{5}$ km)

20.2 Determine the radii of the spheres of influence of Earth, Saturn, and Neptune. (Ans. $r_{SOI_{Earth}} = 9.25\times10^{5}$ km, $r_{SOI_{Saturn}} = 5.48\times10^{7}$ km, $r_{SOI_{Neptune}} = 8.66\times10^{7}$ km)

20.3 Compute the radii of the spheres of influence of Venus and Jupiter.
($r_{SOI_{Venus}} = 6.16 \times 10^5$ km, $r_{SOI_{Jupiter}} = 4.82 \times 10^7$ km)

20.4 Calculate the synodic period between the following planets in years
(a) Mercury and Venus
(b) Pluto and Neptune.
(Ans. (a) $\tau_S = 0.3958$ yr; (b) $\tau_S = 492.412$ yr)

20.5 Find the synodic period between Earth and Venus.
(Ans. $\tau_S = 1.599$ yr)

20.6 Determine the synodic period between the following planets in years
(a) Venus and Mars
(b) Mercury and Earth
(c) Jupiter and Saturn.
(Ans. (a) $\tau_S = 0.914$ yr; (b) $\tau_S = 0.317$ yr; (c) $\tau_S = 19.85$ yr)

20.7 Consider an interplanetary Hohmann transfer from Jupiter to Neptune. Assuming circular orbits for both planets about the Sun, determine the hyperbolic excess speeds at Jupiter departure and at Neptune arrival.
(Ans. $v_{\infty/J}^{+} = 3.99\ \frac{km}{s}$, $v_{\infty/N}^{-} = 2.48\ \frac{km}{s}$)

20.8 Consider an interplanetary Hohmann transfer from Earth to Mercury. Assuming circular orbits for both planets about the Sun, calculate the hyperbolic excess speeds at Earth departure and at Mercury arrival.
(Ans. $v_{\infty/E}^{+} = 7.52\ \frac{km}{s}$, $v_{\infty/M}^{-} = 9.62\ \frac{km}{s}$)

(This page was intentionally left blank.)

# Module 21: Interplanetary Trajectories - Transfer to Superior Planets

The first step in the transfer to a superior planet is to size the heliocentric transfer ellipse, where

- Planet radii can be neglected for sizing the transfer orbit, i.e., the trajectory is designed from the center of the departure planet to the center of the target/arrival planet.

- Hohmann transfers between the planets are used for maximum efficiency.

Heliocentric Transfer Between Planets

Consider a Hohmann transfer to a superior planet where both planets are assumed to be in circular orbits about the Sun. The transfer trajectory is an ellipse sized from the center of the departure planet to the center of the target planet as shown

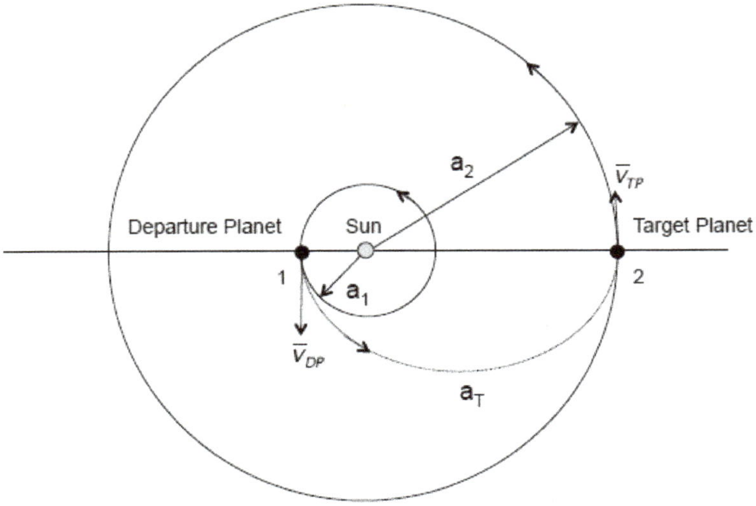


where

$$v_{DP} = \sqrt{\frac{\mu}{a_1}}$$

$$a_T = \frac{a_1+a_2}{2}$$

In transferring to a superior planet, the spacecraft must speed-up relative to the departure planet (to escape its gravitational attraction) and also speed-up relative to the Sun.

$\Rightarrow$ the spacecraft must depart the planet in same direction as the planet's velocity

The velocity change at the departure planet will create the speed required at the periapsis of the transfer ellipse, $\bar{v}_{T1}$, as indicated below

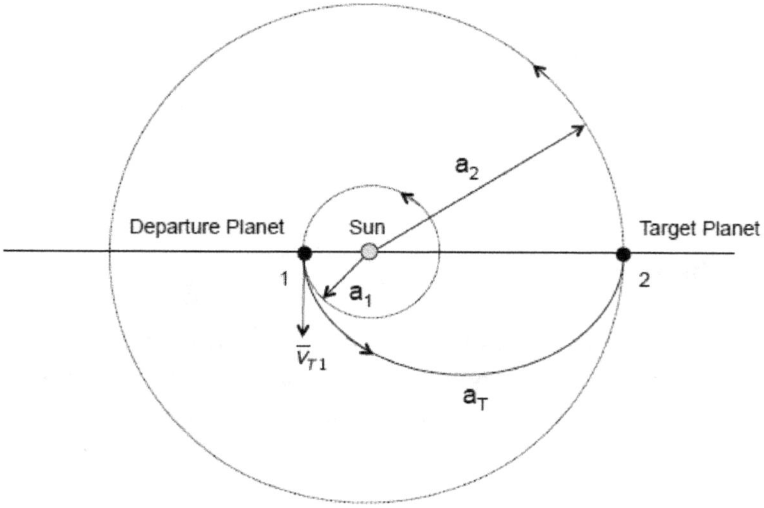


where

$$v_{T1} = \sqrt{\mu\left(\frac{2}{a_1} - \frac{1}{a_T}\right)}$$

The spacecraft will travel along the transfer ellipse and arrive at the apoapsis with the velocity, $\bar{v}_{T2}$, calculated by the expression

$$v_{T2} = \sqrt{\mu\left(\frac{2}{a_2} - \frac{1}{a_T}\right)}$$

Upon arrival, the spacecraft will have less energy than target planet at point 2, therefore the spacecraft cannot catch the planet from behind and must aim ahead of planet and let planet catch the spacecraft from behind

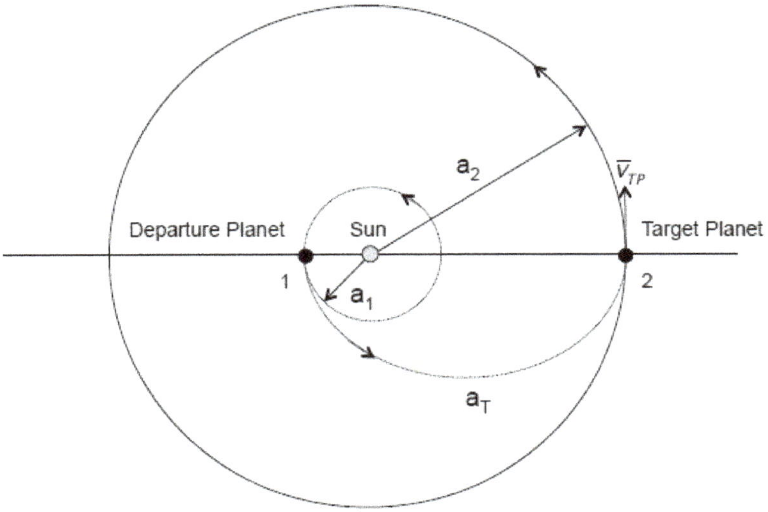


where

$$v_{TP} = \sqrt{\frac{\mu}{a_2}}$$

Planet Departure

Consider the spacecraft to be in a circular orbit having velocity $\bar{v}_c$ relative to the departure planet

A tangential $\Delta v$ is applied at the burnout point in the direction of travel to obtain the burnout velocity, $\bar{v}_{bo}$

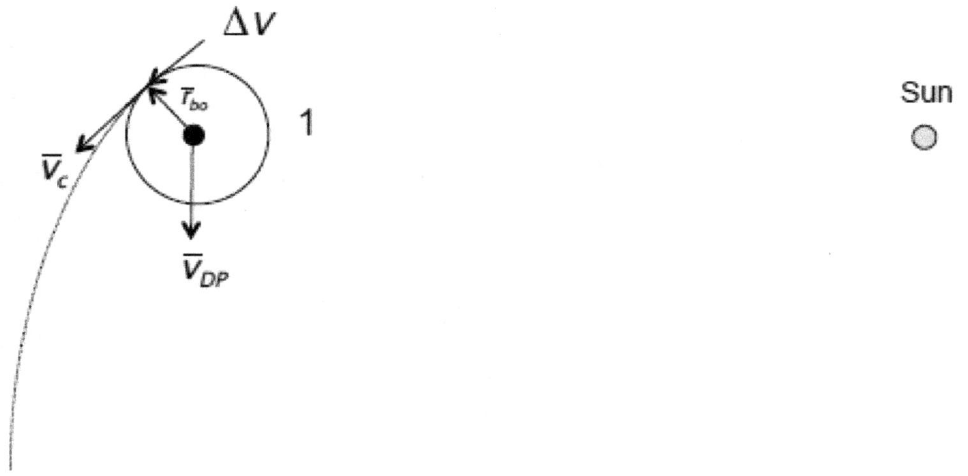


The spacecraft will depart on a hyperbolic trajectory, along orbit 2, which is parallel to the velocity of the departure planet with an initial velocity $\bar{v}_{bo}$ at the burnout point

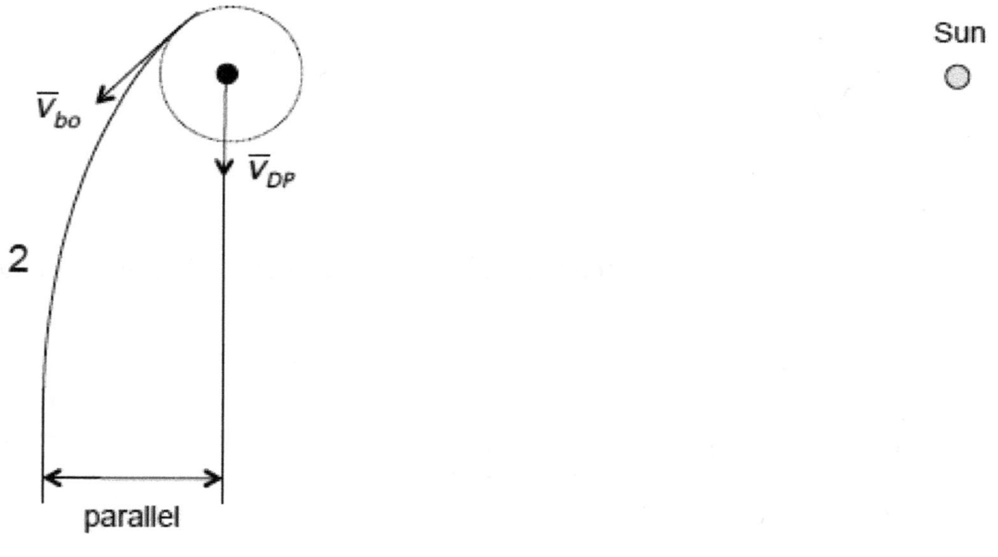

The velocity of the spacecraft decreases until the departure planet's SOI is reached, and the speed remaining is the hyperbolic excess speed at departure, $\bar{v}_{\infty}^{+}$

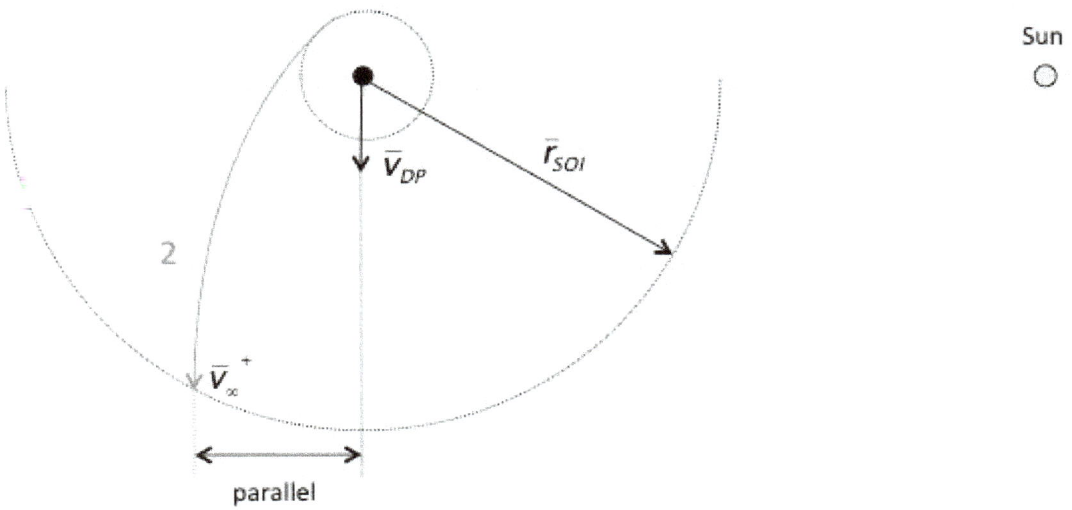


At the SOI, the hyperbolic departure trajectory hands the spacecraft off to the heliocentric elliptical transfer orbit for travel toward the target planet.

The magnitudes of the velocities required for planetary departure are as follows

$$v_{\infty}^{+} = v_{T1} - v_{DP}$$

$$v_{bo} = \sqrt{(v_{\infty}^{+})^2 + \frac{2\mu}{r_{bo}}}$$

$$\Delta v = v_{bo} - v_c$$

Target Planet Arrival

Since the spacecraft has less energy than the target planet, the trajectory is targeted ahead so the planet can catch the spacecraft from behind. When the spacecraft reaches the target planet SOI, the perspective of the spacecraft from the target planet is that the spacecraft is approaching the planet from ahead, with a velocity (relative to the planet) equal to the hyperbolic excess speed, $\bar{v}_{\infty}^{-}$

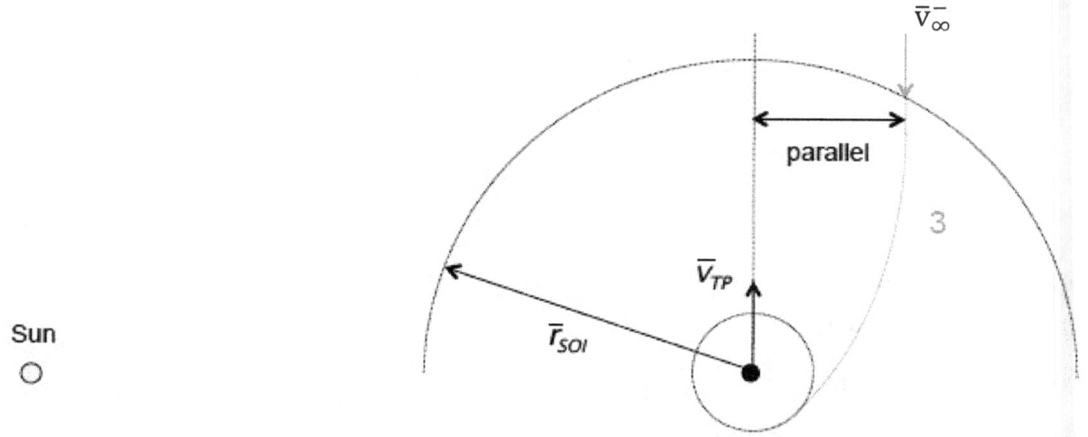


The velocity of the spacecraft increases as it approaches along the asymptote 3, until it reaches the point of closest approach, $\bar{r}_3$, with a velocity $\bar{v}_3$, as shown

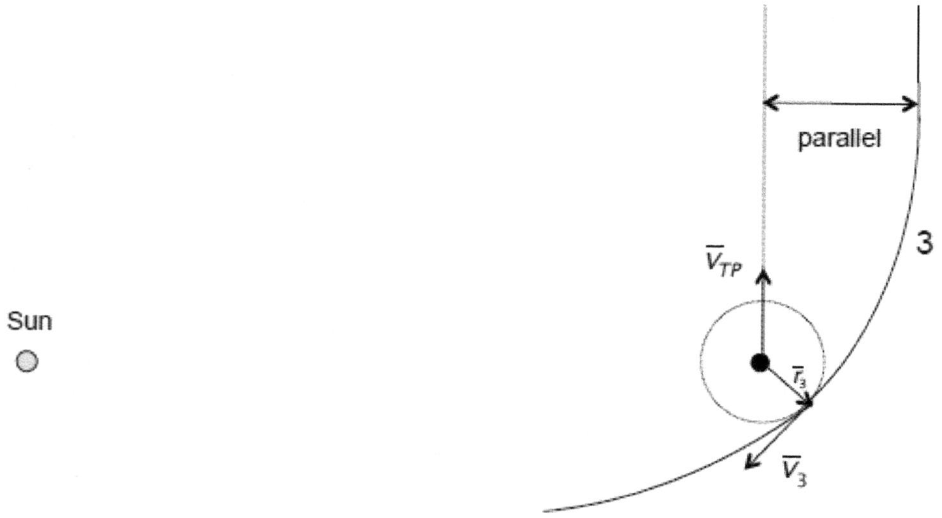

If no additional maneuver is undertaken during the approach trajectory, the spacecraft will depart the target planet on an identical hyperbolic departure trajectory. At the SOI boundary, the spacecraft will have a velocity of $\bar{v}_{\infty}^{+}$. The geometry of the hyperbolic approach/departure trajectories is described in Module 23.

The magnitudes of the velocities at target planet arrival are as follows

$$v_{\infty}^{-} = v_{\infty}^{+}$$

$$v_{\infty}^{-} = v_{TP} - v_{T2}$$

$$v_3 = \sqrt{(v_{\infty}^{-})^2 + \frac{2\mu}{r_3}}$$

Example 21.1

Consider an interplanetary Hohmann transfer from Venus to Saturn. Assume circular orbits of both planets and calculate the hyperbolic excess speeds at Venus departure and at Saturn arrival.

Solution:

$$a_T = \frac{a_V + a_S}{2} = \frac{1.082\times10^8 + 1.433\times10^9}{2} = 7.706 \times 10^8 \text{ km}$$

$$v_V = \sqrt{\frac{\mu_{Sun}}{a_V}} = \sqrt{\frac{1.327\times10^{11}}{1.082\times10^8}} = 35.02 \ \frac{\text{km}}{\text{s}}$$

$$v_{T1} = \sqrt{\mu_{Sun}\left(\frac{2}{a_V} - \frac{1}{a_T}\right)} = \sqrt{(1.327 \times 10^{11})\left(\frac{2}{1.082\times10^8} - \frac{1}{7.706\times10^8}\right)} = 47.76 \ \frac{\text{km}}{\text{s}}$$

$$v_{\infty/V}^{+} = v_{T1} - v_V = 47.76 - 35.02 \ \frac{\text{km}}{\text{s}}$$

$$\Rightarrow \ v_{\infty/V}^{+} = 12.74 \ \frac{\text{km}}{\text{s}}$$

$$v_S = \sqrt{\frac{\mu_{Sun}}{a_S}} = \sqrt{\frac{1.327\times10^{11}}{1.433\times10^9}} = 9.62 \frac{\text{km}}{\text{s}}$$

$$v_{T2} = \sqrt{\mu_{Sun}\left(\frac{2}{a_S} - \frac{1}{a_T}\right)} = \sqrt{(1.327 \times 10^{11})\left(\frac{2}{1.433\times10^9} - \frac{1}{7.706\times10^8}\right)} = 3.61 \ \frac{\text{km}}{\text{s}}$$

$$v_{\infty/S}^{-} = v_S - v_{T2} = 9.62 - 3.61 \ \frac{km}{s}$$

$$\Rightarrow v_{\infty/S}^{-} = 6.01 \frac{km}{s}$$

Example 21.2

A spacecraft is in a 400 km parking orbit about Earth. Determine the velocity change required at Earth to place this spacecraft on a Hohmann transfer trajectory to Uranus.

Solution:

$$a_T = \frac{a_E + a_U}{2} = \frac{1.496\times10^8 + 2.872\times10^9}{2} = 1.511 \times 10^9 \text{ km}$$

$$v_{E/Sun} = \sqrt{\frac{\mu_{Sun}}{a_E}} = \sqrt{\frac{1.327\times10^{11}}{1.496\times10^8}} = 29.78 \ \frac{km}{s}$$

$$v_{T1} = \sqrt{\mu_{Sun}\left(\frac{2}{a_E} - \frac{1}{a_T}\right)} = \sqrt{(1.327 \times 10^{11})\left(\frac{2}{1.496\times10^8} - \frac{1}{1.511\times10^9}\right)} = 41.06 \ \frac{km}{s}$$

$$v_{\infty/E}^{+} = v_{T1} - v_{E/Sun} = 41.06 - 29.78 \ \frac{km}{s} = 11.28 \ \frac{km}{s}$$

$$r_E = R_E + 400 = 6{,}378.1 + 400 = 6{,}778.1 \text{ km}$$

$$v_{bo} = \sqrt{(v_{\infty}^{+})^2 + \frac{2\mu_E}{r_E}} = \sqrt{(11.28)^2 + \frac{2(3.986\times10^5)}{6{,}778.1}} = 15.65 \ \frac{km}{s}$$

$$v_{SC/E} = \sqrt{\frac{\mu_E}{r_E}} = \sqrt{\frac{3.986\times10^5}{6{,}778.1}} = 7.67 \ \frac{km}{s}$$

$$\Delta v = v_{bo} - v_{SC/E} = 15.65 - 7.67 \ \frac{km}{s}$$

$$\Rightarrow \Delta v = 7.98 \ \frac{km}{s}$$

Example 21.3

For Example 21.2, find the speed of the spacecraft at its closest approach to Uranus at an altitude of 500 km.

Solution

$$a_T = \frac{a_E + a_U}{2} = \frac{1.496\times10^8 + 2.872\times10^9}{2} = 1.511 \times 10^9 \text{ km}$$

$$v_{J/Sun} = \sqrt{\frac{\mu_{Sun}}{a_U}} = \sqrt{\frac{1.327\times10^{11}}{2.872\times10^9}} = 6.80\ \frac{\text{km}}{\text{s}}$$

$$v_{T2} = \sqrt{\mu_{Sun}\left(\frac{2}{a_U} - \frac{1}{a_T}\right)} = \sqrt{(1.327 \times 10^{11})\left(\frac{2}{2.872\times10^9} - \frac{1}{1.511\times10^9}\right)} = 2.14\ \frac{\text{km}}{\text{s}}$$

$$v^-_{\infty/U} = v_{U/Sun} - v_{T2} = 6.80 - 2.14\ \frac{\text{km}}{\text{s}} = 4.66\ \frac{\text{km}}{\text{s}}$$

$$r_J = R_U + (\text{alt})_U = 25{,}560 + 500 = 30{,}060 \text{ km}$$

$$v_3 = \sqrt{(v^-_\infty)^2 + \frac{2\mu_U}{r_U}} = \sqrt{(4.66)^2 + \frac{2(5.794\times10^6)}{30{,}060}} = 20.18\ \frac{\text{km}}{\text{s}}$$

Problems

21.1 Consider an interplanetary Hohmann transfer from Mars to Saturn. Assuming circular orbits of both planets about the Sun, calculate the hyperbolic excess speeds at Mars departure and Saturn arrival.
(Ans. $v^+_{\infty/M} = 7.57\ \frac{\text{km}}{\text{s}}$, $v^-_{\infty/S} = 4.58\ \frac{\text{km}}{\text{s}}$)

21.2 If the spacecraft in Problem 21.1 is to leave Mars from a 700 km circular orbit, determine the $\Delta v$ required to achieve the desired hyperbolic excess speed at Mars departure.
(Ans. $\Delta v = 5.61\ \frac{\text{km}}{\text{s}}$)

21.3 For the mission described in Problems 21.1 and 21.2, find the speed of the spacecraft at its point of closest approach to Saturn at an altitude of 1,000 km.
(Ans. $v = 35.47\ \frac{\text{km}}{\text{s}}$)

21.4 Consider an interplanetary Hohmann transfer from Mercury to Venus. Assuming circular orbits of both planets about the Sun, calculate the hyperbolic excess speeds at Mercury departure and Venus arrival.

(Ans. $v^+_{\infty/M} = 6.77\ \frac{km}{s}$, $v^-_{\infty/V} = 5.78\ \frac{km}{s}$)

21.5 If the spacecraft in Problem 21.4 is to leave Mercury from a 450 km circular orbit, determine the $\Delta v$ required to achieve the desired hyperbolic excess speed at Mercury departure.

(Ans. $\Delta v = 5.05\ \frac{km}{s}$)

21.6 For the mission described in Problems 21.4 and 21.5, find the speed of the spacecraft at its point of closest approach to Venus at an altitude of 300 km.

(Ans. $v = 11.65\ \frac{km}{s}$)

21.7 Consider an interplanetary Hohmann transfer from Earth to Pluto. Assuming circular orbits of both planets about the Sun, calculate the hyperbolic excess speeds at Earth departure and Pluto arrival.

(Ans. $v^+_{\infty/M} = 11.81\ \frac{km}{s}$, $v^-_{\infty/V} = 3.69\frac{km}{s}$)

21.8 If the spacecraft in Problem 21.7 is to leave Earth from a 300 km circular orbit, Determine

(a) the $\Delta v$ required to achieve the desired hyperbolic excess speed at Earth departure

(b) speed of the spacecraft at its point of closest approach at Pluto at an altitude of 200 km

(c) the approximate flight time from Earth departure to Pluto arrival in Earth years.

(Ans. (a) $\Delta v = 8.36\ \frac{km}{s}$; (b) $v = 3.85\ \frac{km}{s}$; (c) $t = 45.13$ yr)

# Module 22: Interplanetary Trajectories - Transfer to Inferior Planets

The first step in a transfer to an inferior planet is also to size the heliocentric transfer orbit (ellipse), where again

- Planet radii can be neglected for sizing the transfer orbit, i.e., the trajectory is designed from the center of the departure planet to the center of the target/arrival planet.

- Hohmann transfers between the planets are used for maximum efficiency.

<u>Heliocentric Transfer Between Planets</u>

Consider a Hohmann transfer to an inferior planet where both planets are in circular orbits about the Sun. The transfer trajectory is an ellipse sized from the departure planet to the target planet as shown

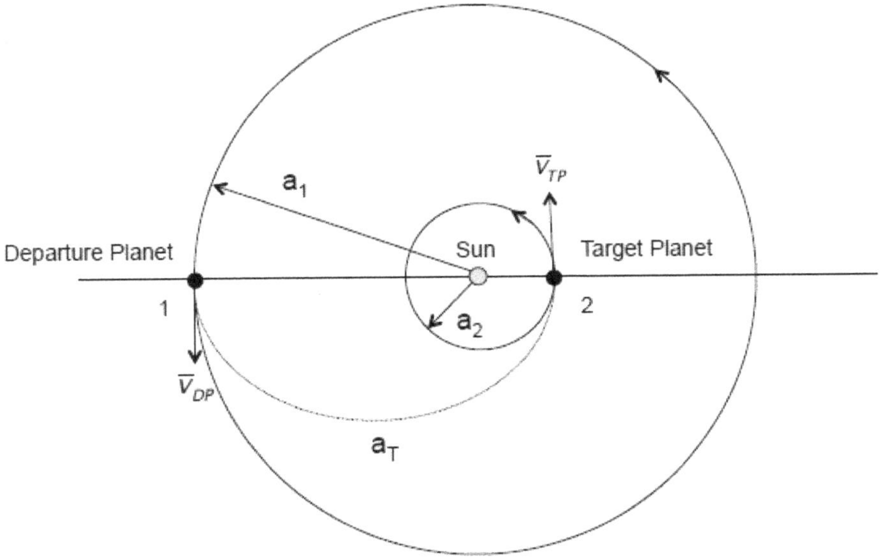


The spacecraft must speed-up relative to departure planet (to escape its gravitational attraction) but slow-down relative to the Sun at point 1.

$\Rightarrow$ the spacecraft must depart the planet in the opposite direction of the planet's velocity.

The velocity change will create the speed required at the apoapsis of the transfer ellipse, $\bar{v}_{T1}$, as indicated below

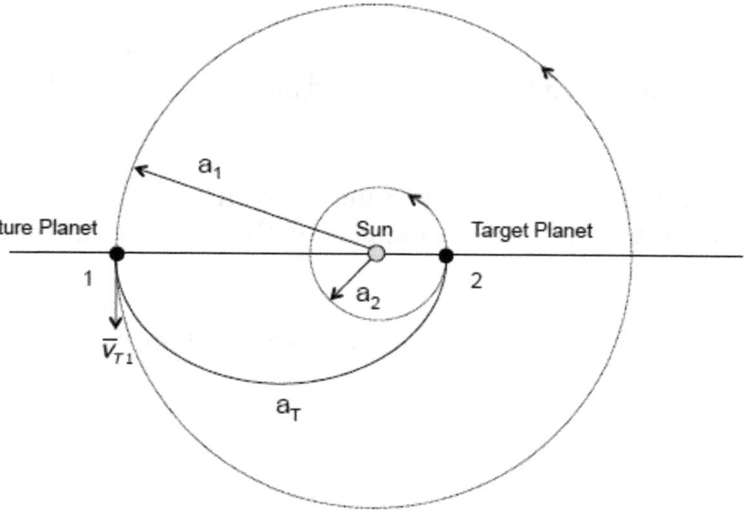


The spacecraft will travel along the transfer ellipse and arrive at the periapsis with the velocity, $\bar{v}_{T2}$.

Upon arrival, the spacecraft will have more energy than target planet at point 2, so the spacecraft can catch the planet from behind.

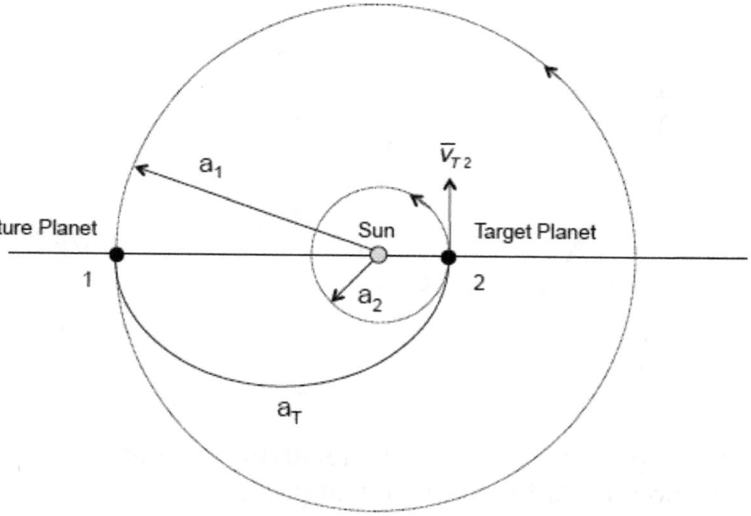

Planet Departure

The spacecraft must speed-up relative to departure planet (to escape) but slow down relative to the Sun.

$\Rightarrow$ spacecraft must depart planet in the direction opposite to planet's velocity.

The spacecraft is in a circular orbit having speed $\bar{v}_c$ relative to the departure planet

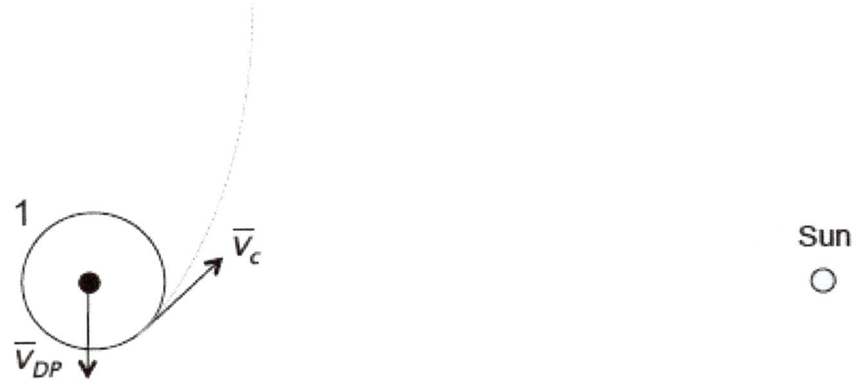


A tangential $\Delta\bar{v}$ is applied at the burnout point in the direction of travel to obtain the burnout velocity,$\bar{v}_{bo}$ and move onto the hyperbolic departure trajectory

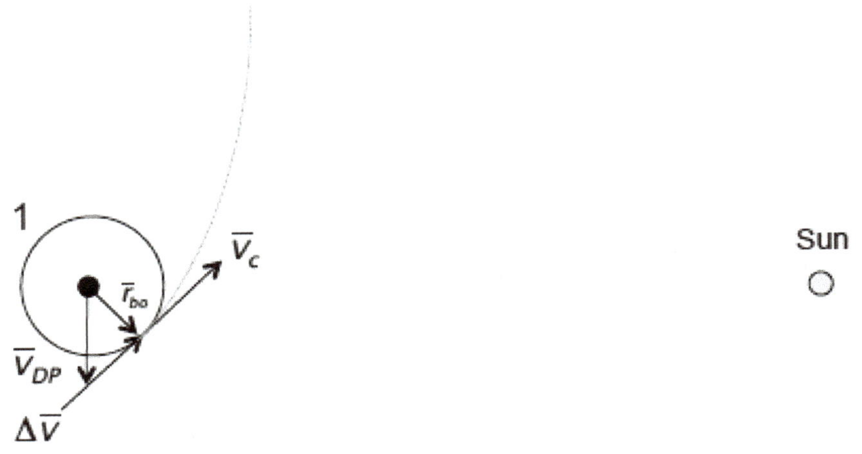

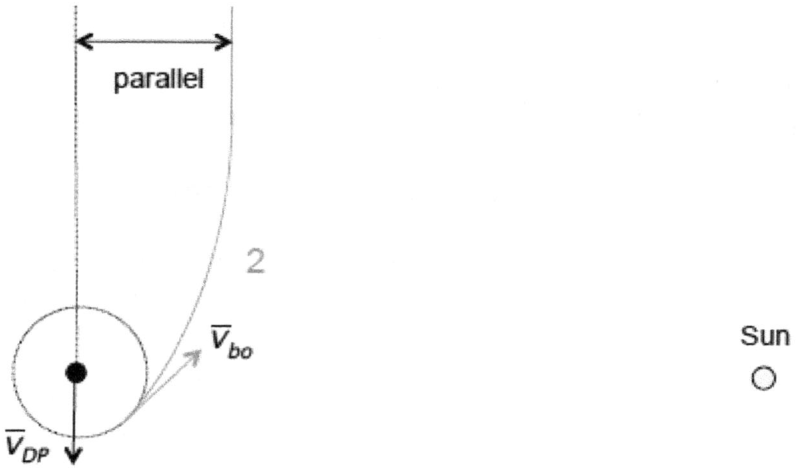


The velocity will decrease until the departure planet's SOI is reached, and the speed remaining is the hyperbolic excess speed at departure, $\bar{v}_\infty^+$

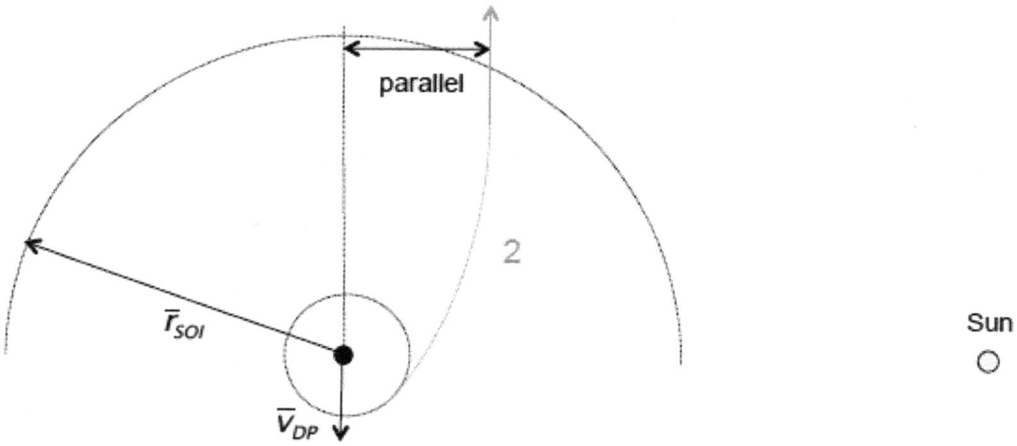


At the SOI, the hyperbolic departure trajectory hands the spacecraft off to the heliocentric elliptical transfer orbit for travel toward the target planet.

The magnitudes of the velocities required for planetary departure are as follows

$$v_\infty^+ = v_{DP} - v_{T1}$$

$$v_{bo} = \sqrt{(v_\infty^+)^2 + \frac{2\mu}{r_{bo}}}$$

$$\Delta v = v_{bo} - v_c$$

Target Planet Arrival

Since the spacecraft has more energy than the target planet, the spacecraft will catch the target planet from behind. When the spacecraft reaches the target planet's SOI, the perspective from target planet is that the spacecraft is approaching from behind, with a velocity (relative to the planet) equal to the hyperbolic excess speed, $\bar{v}_\infty^-$

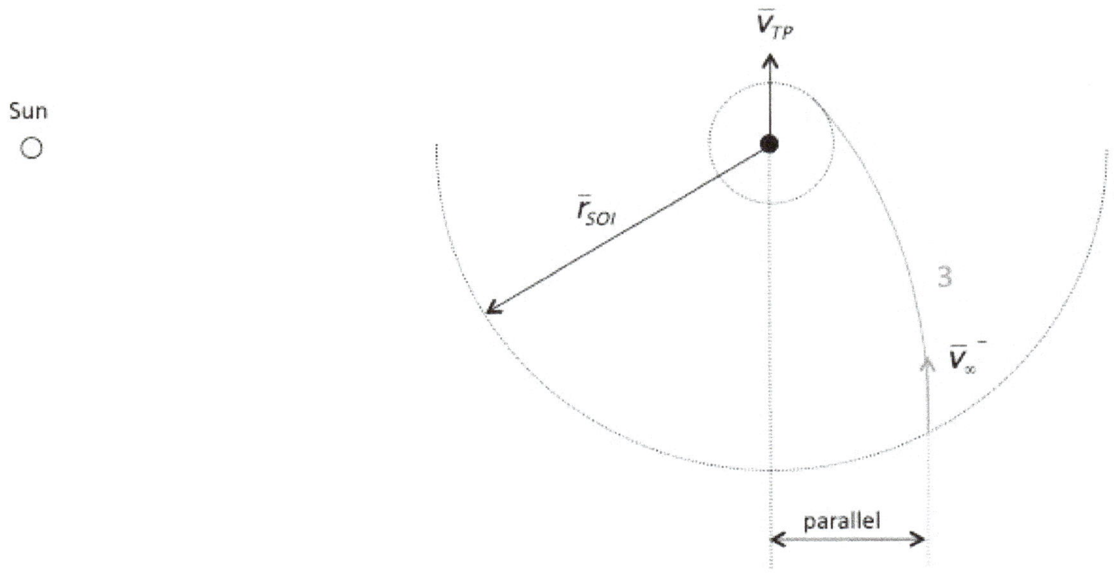


The velocity of the spacecraft increases as it approaches along the asymptote 3, until it reaches the point of closest approach, $\bar{r}_3$, with a velocity $\bar{v}_3$, as shown

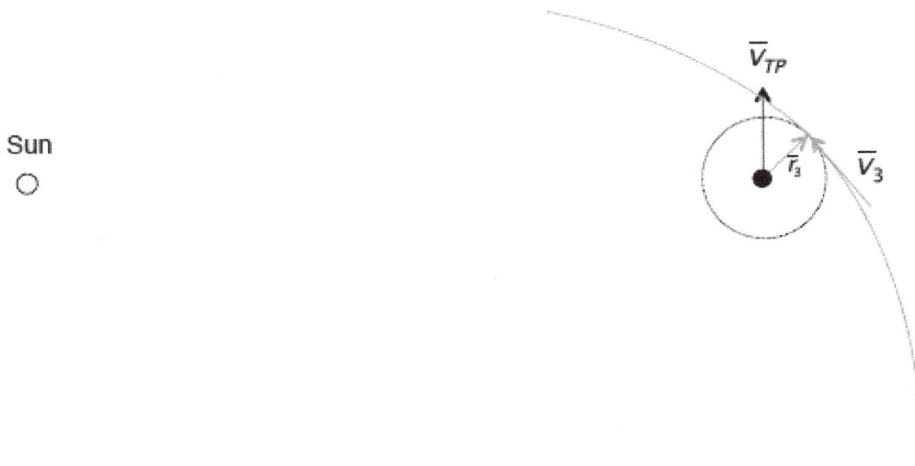

If no additional maneuver is undertaken during the approach trajectory, the spacecraft will depart the target planet on an identical hyperbolic departure trajectory. At the SOI boundary, the spacecraft will have a velocity of $\bar{v}_{\infty}^{+}$. The geometry of the hyperbolic approach/departure trajectories is described in Module 23.

The magnitudes of the velocities at the target planet arrival are as follows

$$v_{\infty}^{-} = v_{\infty}^{+}$$

$$v_{\infty}^{-} = v_{T2} - v_{TP}$$

$$v_3 = \sqrt{(v_{\infty}^{-})^2 + \frac{2\mu}{r_3}}$$

Example 22.1

Consider an interplanetary Hohmann transfer from Neptune to Saturn. Assume circular orbits of both planets about the Sun. Calculate the hyperbolic excess speeds at Neptune departure and at Saturn arrival.

Solution:

$$a_T = \frac{a_N + a_S}{2} = \frac{4.495\times10^9 + 1.433\times10^9}{2} = 2.964\times10^8\ \text{km}$$

$$v_N = \sqrt{\frac{\mu_{Sun}}{a_N}} = \sqrt{\frac{1.327\times10^{11}}{4.495\times10^9}} = 5.43\ \frac{\text{km}}{\text{s}}$$

$$v_S = \sqrt{\frac{\mu_{Sun}}{a_S}} = \sqrt{\frac{1.327\times10^{11}}{1.433\times10^9}} = 9.62\frac{\text{km}}{\text{s}}$$

$$v_{T1} = \sqrt{\mu_{Sun}\left(\frac{2}{a_N} - \frac{1}{a_T}\right)} = \sqrt{(1.327\times10^{11})\left(\frac{2}{4.495\times10^9} - \frac{1}{2.964\times10^9}\right)} = 3.78\ \frac{\text{km}}{\text{s}}$$

$$v_{T2} = \sqrt{\mu_{Sun}\left(\frac{2}{a_S} - \frac{1}{a_T}\right)} = \sqrt{(1.327\times10^{11})\left(\frac{2}{1.433\times10^9} - \frac{1}{2.964\times10^9}\right)} = 11.85\ \frac{\text{km}}{\text{s}}$$

$$v_{\infty/N}^{+} = v_N - v_{T1} = 5.43 - 3.78\ \frac{\text{km}}{\text{s}}$$

$$\Rightarrow\ v_{\infty/N}^{+} = 1.65\ \frac{\text{km}}{\text{s}}$$

$$v_{\infty/S}^{-} = v_{T2} - v_S = 11.85 - 9.62 \ \frac{km}{s}$$

$$\Rightarrow \ v_{\infty/S}^{-} = 2.23 \ \frac{km}{s}$$

Example 22.2

Consider an interplanetary Hohmann transfer from Saturn to Mars. Assume circular orbits of both planets about the Sun. Determine the hyperbolic excess speeds at Saturn departure and at Mars arrival.

Solution

$$a_T = \frac{a_S + a_M}{2} = \frac{1.433\times10^{9} + 2.279\times10^{8}}{2} = 8.305 \times 10^{8} \text{ km}$$

$$v_S = \sqrt{\frac{\mu_{Sun}}{a_S}} = \sqrt{\frac{1.327\times10^{11}}{1.433\times10^{9}}} = 9.62 \frac{km}{s}$$

$$v_M = \sqrt{\frac{\mu_{Sun}}{a_N}} = \sqrt{\frac{1.327\times10^{11}}{2.279\times10^{9}}} = 24.13 \ \frac{km}{s}$$

$$v_{T1} = \sqrt{\mu_{Sun}\left(\frac{2}{a_S} - \frac{1}{a_T}\right)} = \sqrt{(1.327 \times 10^{11})\left(\frac{2}{1.433\times10^{9}} - \frac{1}{8.305\times10^{8}}\right)} = 5.04 \ \frac{km}{s}$$

$$v_{T2} = \sqrt{\mu_{Sun}\left(\frac{2}{a_M} - \frac{1}{a_T}\right)} = \sqrt{(1.327 \times 10^{11})\left(\frac{2}{2.279\times10^{8}} - \frac{1}{8.305\times10^{8}}\right)} = 31.70 \ \frac{km}{s}$$

$$v_{\infty/S}^{+} = v_S - v_{T1} = 9.62 - 5.04 \ \frac{km}{s}$$

$$\Rightarrow \ v_{\infty/S}^{+} = 4.58 \ \frac{km}{s}$$

$$v_{\infty/M}^{-} = v_{T2} - v_M = 31.70 - 24.13 \ \frac{km}{s}$$

$$\Rightarrow \ v_{\infty/M}^{-} = 7.57 \ \frac{km}{s}$$

Example 22.3

For Example 22.2, find the Δv required for a spacecraft to depart Saturn from a circular orbit at an altitude of 1,000 km.

Solution:

$$r_S = R_S + (\text{alt})_{Sat} = 60{,}270 + 1{,}000 = 61{,}270 \text{ km}$$

$$v_{SC/S} = \sqrt{\frac{\mu_S}{r_{SC/S}}} = \sqrt{\frac{3.793\times10^7}{61{,}270}} = 24.88\ \frac{\text{km}}{\text{s}}$$

$$v_{bo} = \sqrt{(v_\infty^+)^2 + \frac{2\mu_S}{r_S}} = \sqrt{(4.58)^2 + \frac{2(3.793\times10^7)}{61{,}270}} = 35.50\ \frac{\text{km}}{\text{s}}$$

$$\Delta v = v_{bo} - v_{SC/S} = 35.50 - 24.88$$

$$\Rightarrow\ \Delta v = 10.62\frac{\text{km}}{\text{s}}$$

Problems

22.1 For an interplanetary Hohmann transfer from Earth to Venus, calculate the hyperbolic excess speeds at Earth departure and Venus arrival. Assume circular orbits of both planets about the Sun,
(Ans. $v_{\infty/E}^+ = 2.50\ \frac{\text{km}}{\text{s}}, v_{\infty/V}^- = 2.71\ \frac{\text{km}}{\text{s}}$)

22.2 If the spacecraft in Problem 22.1 is to leave Earth from a 600 km circular orbit, determine the Δv required to achieve the desired hyperbolic excess speed at Earth departure.
(Ans. $\Delta v = 3.42\ \frac{\text{km}}{\text{s}}$)

22.3 For the mission described in Problems 22.1 and 22.2, find the speed of the spacecraft at its point of closest approach to Venus at an altitude of 500 km.
(Ans. $v = 10.32\frac{\text{km}}{\text{s}}$)

22.4 For an interplanetary Hohmann transfer from Saturn to Venus, calculate the hyperbolic excess speeds at Saturn departure and Venus arrival. Assume circular orbits of both planets about the Sun,
(Ans. $v_{\infty/S}^+ = 6.02\ \frac{\text{km}}{\text{s}}, v_{\infty/V}^- = 12.74\ \frac{\text{km}}{\text{s}}$)

22.5 If the spacecraft in Problem 22.4 is to leave Saturn from a 1,200 km circular orbit, determine the Δv required to achieve the desired hyperbolic excess speed at Saturn departure.

(Ans. $\Delta v = 10.80\ \frac{km}{s}$)

22.6 For the mission described in Problems 22.4 and 22.5, find the speed of the spacecraft at its point of closest approach to Venus at an altitude of 425 km.

(Ans. $v = 7.08\ \frac{km}{s}$)

22.7 Consider an interplanetary Hohmann transfer from Jupiter to Earth. Assuming circular orbits of both planets about the Sun, calculate the hyperbolic excess speeds at Jupiter departure and Earth arrival.

(Ans. $v^{+}_{\infty/J} = 5.65\ \frac{km}{s}$, $v^{-}_{\infty/E} = 8.80\frac{km}{s}$)

22.8 If the spacecraft in Problem 22.7 is to leave Jupiter from a 1,500 km circular orbit, Determine

(a) the Δv required to achieve the desired hyperbolic excess speed at Jupiter departure

(b) speed of the spacecraft at its point of closest approach at Earth at an altitude of 250 km.

(c) the approximate flight time from Jupiter departure to Earth arrival in Earth years.

(Ans. (a) $\Delta v = 17.53\ \frac{km}{s}$; (b) $v = 14.06\ \frac{km}{s}$; (c) t = 2.73 yr)

(This page was intentionally left blank.)

# Module 23: Interplanetary Trajectories - Hyperbolic Passage and Planetary Flybys

## Hyperbolic Passage

The geometry of hyperbolic approach/departure trajectories is referred to as a hyperbolic passage. Consider the geometry of the approach and departure trajectories in the following figure

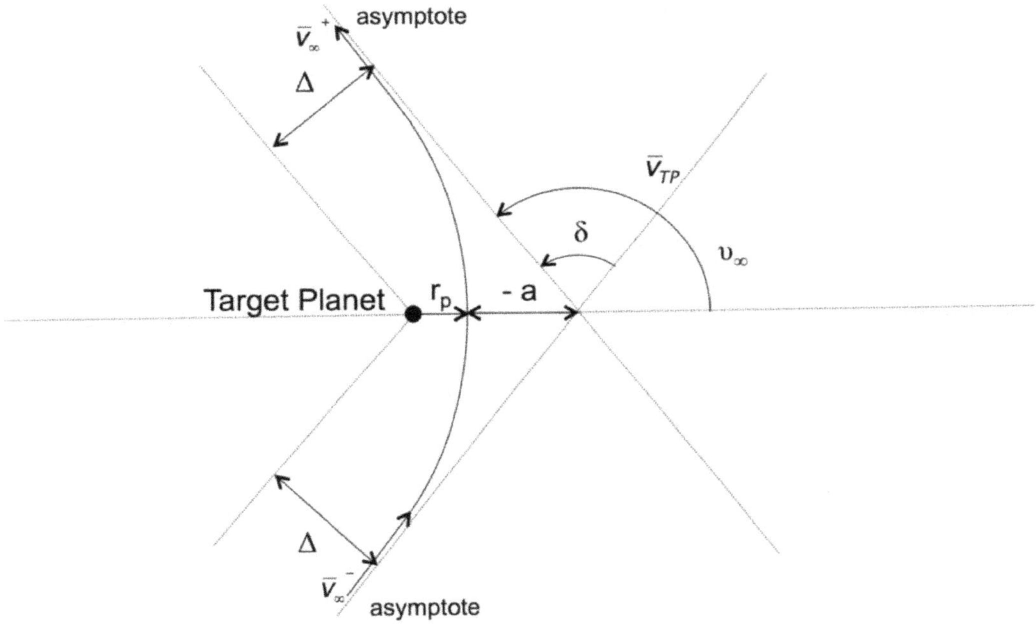


where

$r_p$ = distance to periapsis
$\Delta$ = approach/aiming distance
$\delta$ = turning angle
$\upsilon_\infty$ = true anomaly of the asymptote
$v_\infty^- = |\bar{v}_\infty^-|$, hyperbolic excess speed at approach
$v_\infty^+ = |\bar{v}_\infty^+|$, hyperbolic excess speed at departure

Because the total energy of $m_2$ is constant $v_\infty^+ = v_\infty^- = v_\infty$

$\Rightarrow$ the hyperbolic excess speeds have equal magnitudes but different directions

In designing a hyperbolic passage of a planet, usually $v_\infty$ and either $r_p$ or $\Delta$ are known. The other quantities are related by the following equations

$$a = \frac{-\mu}{v_\infty^2}$$

$$v_\infty = \cos^{-1}\left(\frac{-1}{e}\right)$$

$$\delta = 2\,\sin^{-1}\left(\frac{1}{e}\right)$$

$$e = 1 + \frac{r_p v_\infty^2}{\mu} = \sqrt{1 + \frac{V_\infty^4 \Delta^2}{\mu^2}}$$

$$\Delta = r_p\sqrt{1 + \frac{2\mu}{r_p v_\infty^2}} = \sqrt{\frac{\mu^2(e^2-1)}{v_\infty^4}}$$

$$r_p = a + \frac{\Delta}{\cos\left(\frac{\delta}{2}\right)}$$

When a spacecraft approaches a planet, three outcomes are possible:

1. The spacecraft can perform a flyby of the planet, resulting in a gain, loss, or no change in the energy of the spacecraft relative to the Sun.

2. The spacecraft can collide with the planet, which is not usually the desired result.

3. The spacecraft can be captured by the planet, which can occur due to a reduction in the energy of the spacecraft relative to the planet.

Outcome 3 is discussed in the Module 24.

## Example 23.1

On a Hohmann transfer from an inferior planet, a spacecraft approaches Uranus with a hyperbolic excess speed of 4.0 $\frac{\text{km}}{\text{s}}$ on the sunlit side. For the hyperbolic passage, calculate the eccentricity, semi-major axis, turning angle, and aiming distance. The radius of closest approach at Uranus is 25,500 km.

Solution:

$$e = \frac{r_p v_\infty^2}{\mu_U} + 1 = \frac{(25{,}500)(4.0)^2}{5.974\times10^6} + 1$$

$$\Rightarrow \quad e = 1.068$$

$$r_p = a(1 - e)$$

$$a = \frac{r_p}{1-e} = \frac{25{,}500}{1-1.068}$$

$$\Rightarrow \quad a = -375{,}000 \text{ km}$$

$$\delta = 2\sin^{-1}\left(\frac{1}{e}\right) = 2\sin^{-1}\left(\frac{1}{1.068}\right)$$

$$\Rightarrow \quad \delta = 138.9^{\text{o}}$$

$$\Delta = \sqrt{\frac{\mu_U^2(e^2-1)}{v_\infty^4}} = \sqrt{\frac{(5.974\times10^6)^2[(1.068)^2-1]}{(4.0)^4}}$$

$$\Rightarrow \quad \Delta = 140{,}015.1 \text{ km}$$

## Planetary Flybys

Planetary flybys can be designed to either increase or decrease the energy of a spacecraft relative to the Sun during the planetary encounter. A flyby can also result in no change in spacecraft energy.

While the velocity of the spacecraft relative to the planet on approach and departure is constant in magnitude, the inertial velocity, i.e., relative to the Sun, is not necessarily constant in magnitude or direction.

The geometry of the hyperbolic passage can be designed to change inertial velocity to satisfy mission objectives, e.g., return home, or travel to other planets. The effect of planetary flybys on the inertial velocity of the spacecraft is best demonstrated by the concept of relative velocity. Consider the vector addition

$$\bar{v}_{SC/S} = \bar{v}_{P/S} + \bar{v}_{SC/P} \tag{18}$$

This equation reads as

(the velocity of the spacecraft relative to the Sun) = (velocity of the planet relative to the Sun) + (the velocity of the spacecraft relative to the planet)

On approach and departure

$$\bar{v}_{SC/P} = \bar{v}_{\infty}$$

Consider the hyperbolic passage geometry in the following figure

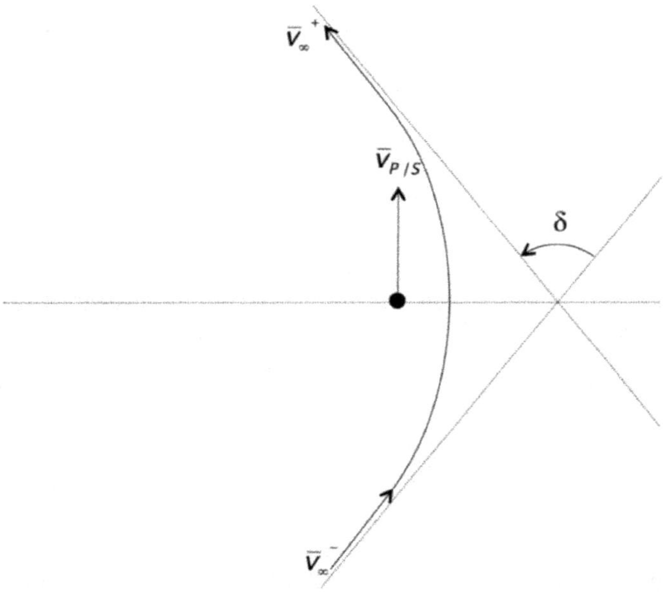


The turning angle, δ, which is a function of the geometry of the hyperbolic passage, is the most important factor in determining whether a spacecraft will gain or lose energy during a flyby.

Consider the graphical representation of Equation (18) at 'planet approach' and 'planet departure' shown below. The <u>net effect</u> of the planetary encounter on the spacecraft inertial velocity is a CCW rotation through angle θ, a result of the size of the turning angle

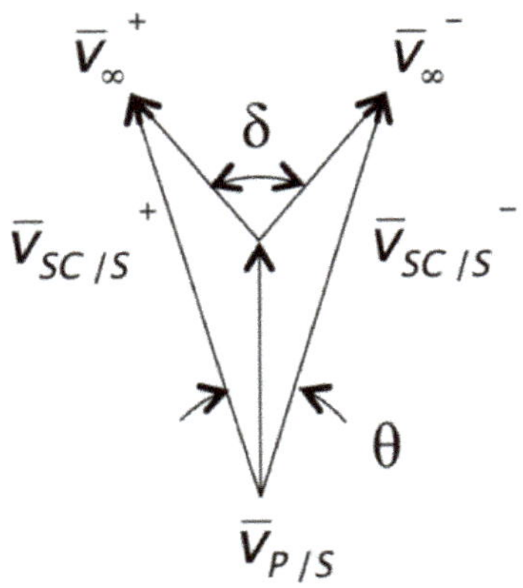


Due to the nature of this passage, it can be seen that

$$|\bar{v}_{SC/S}^+| = |\bar{v}_{SC/S}^-|$$

Therefore, no energy is gained or lost during this passage.

The approach and departure trajectories of a flyby can be designed to change both the direction and magnitude of a spacecraft's inertial velocity through an increase or decrease in orbital energy.

If a spacecraft passes in front of the planet, i.e., front passage, the spacecraft will lose energy.

If a spacecraft passes behind the planet, i.e., back passage, the spacecraft will gain energy.

Both front and back passages are shown in the figures below.

Front Passage

In the front passage shown, the turning angle is greater than 90° in the clockwise direction

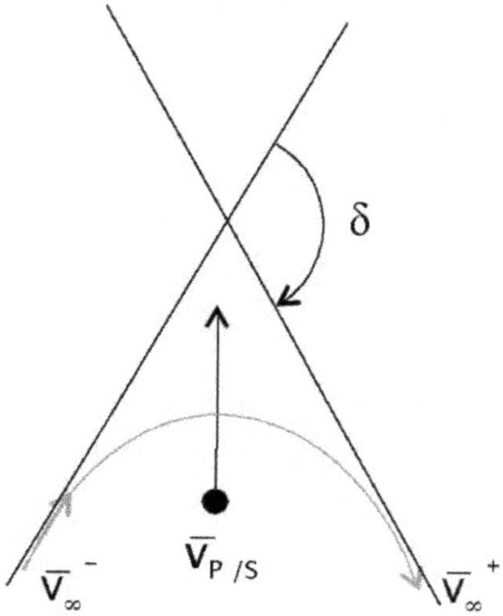


The result of the front passage is a large decrease in the magnitude of the inertial velocity relative to the Sun, $\bar{v}_{SC/S}$, in addition to a clockwise rotation of its direction through angle θ, as shown

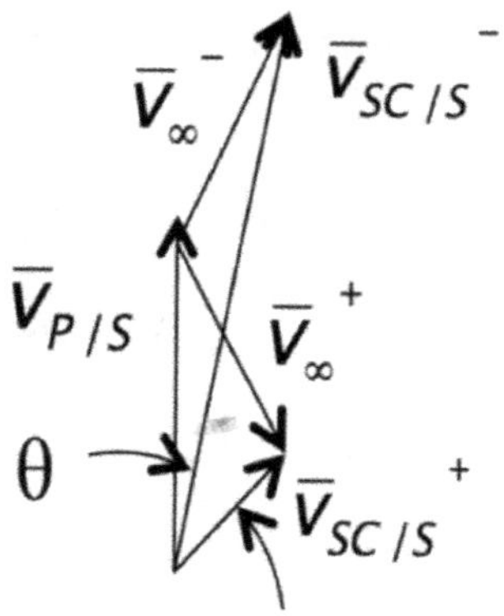

Back Passage

In the back passage shown, the turning angle is greater than 90° in the CCW direction

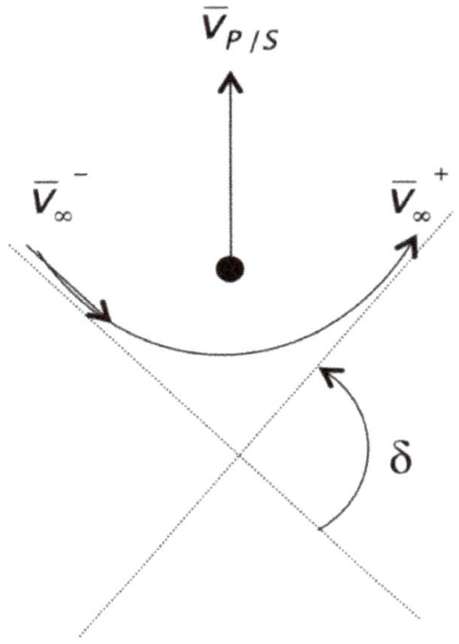


The result of the front passage is a large increase in the magnitude of the inertial velocity relative to the Sun, $\bar{v}_{SC/S}$, in addition to a CCW rotation of its direction through angle θ, as shown

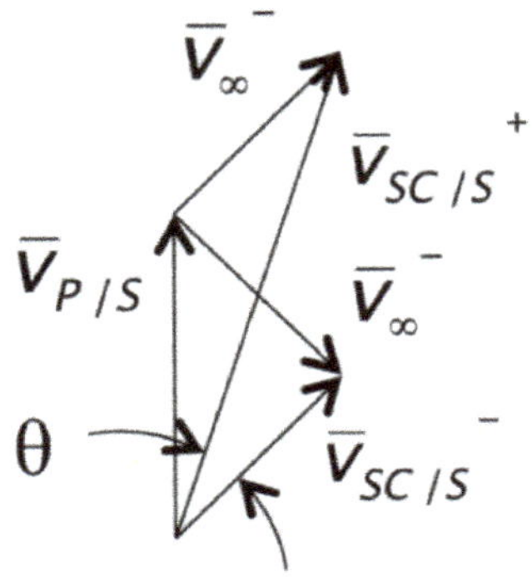


Creative design of the hyperbolic passage can generate an inertial velocity of the spacecraft required for many possible missions after flyby of the target planet.

Example 23.2

A spacecraft is approaching Jupiter with a hyperbolic excess speed of 5.0 $\frac{\text{km}}{\text{s}}$. During a back passage on the sunlit side, the spacecraft achieves a turning angle of $100^o$. Assuming Jupiter has a circular orbit about the Sun, calculate the eccentricity, aiming distance, semi-major axis, and the distance of closest approach of the spacecraft during this passage

Solution:

$$e_J = \frac{1}{\sin\left(\frac{\delta_J}{2}\right)} = \frac{1}{\sin\left(\frac{100^o}{2}\right)}$$

$$\Rightarrow \ e_J = 1.305$$

$$\Delta_J = \sqrt{\frac{\mu_J^2\left(e_J^2-1\right)}{v_\infty^4}} = \sqrt{\frac{(1.267\times10^8)^2[(1.305)^2-1]}{(5.0)^4}}$$

$$\Rightarrow \ \Delta_J = 4{,}249{,}345.0 \text{ km}$$

$$a_J = \frac{-\mu_J}{v_\infty^2} = \frac{-1.267\times10^8}{(5.0)^2}$$

$$\Rightarrow \ a_J = -5{,}068{,}000.0 \text{ km}$$

$$r_p = a_J(1-e_J) = -5{,}068{,}000.0(1-1.305)$$

$$\Rightarrow \ r_p = 1{,}545{,}740.0 \text{ km}$$

Example 23.3

In Example 23.2, find the magnitude of the velocity of the spacecraft relative to the Sun after passage of Jupiter.

Solution:

$$v_{J/Sun} = \sqrt{\frac{\mu_{Sun}}{a_J}} = \sqrt{\frac{1.327\times10^{11}}{7.78\times10^8}} = 13.06\,\frac{\text{km}}{\text{s}}$$

$$v^{+}_{SC/Sun} = \sqrt{v^2_{J/Sun} + \left(v^{+}_{SC/Sun}\right)^2 - 2v_{J/Sun}\left(v^{+}_{SC/Sun}\right)\cos\delta_J}$$

$$v^{+}_{SC/Sun} = \sqrt{(13.06)^2 + (5.0)^2 - 2(13.06)(5.0)\cos 100^o}$$

$$\Rightarrow \quad v^{+}_{SC/Sun} = 14.77\ \frac{km}{s}$$

<u>Problems</u>

23.1 A spacecraft is approaching Saturn with a hyperbolic excess speed of 5.64 $\frac{km}{s}$. During a back passage, the spacecraft achieves a turning angle of 105°. Assume that Saturn has a circular orbit about the Sun and compute the velocity vector of the spacecraft relative to the Sun after passage.
(Ans. $\bar{v}_{SC/Sun} = 5.45\,\bar{I} + 11.09\,\bar{J}\ \frac{km}{s}$)

23.2 In transferring to a superior planet using a Hohmann transfer, a spacecraft is approaching Planet X on the sunlit side with a hyperbolic excess speed of 6.0 $\frac{km}{s}$. During a back passage, the turning angle is 110°. Assume a circular orbit for Planet X having $a_X = 8.0 \times 10^8$ km. Calculate the magnitude of the resulting inertial velocity of the spacecraft after passage and its angle of orientation relative to the velocity vector of Planet X.
(Ans. $v_{X/Sun} = 12.88\ \frac{km}{s}, \theta = 20.687^o$)

23.3 A 15,000 kg space probe is parked in a 600 km direct, circular Earth orbit and must transfer to an orbit which will pass within 1,000 km of the surface of Uranus on the dark side. Use a Hohmann transfer from Earth to Uranus, consider all orbits to lie in the ecliptic plane, and assume that the orbits of Earth and Uranus about the Sun are circular. All velocity changes can be considered to be impulsive and perform the following tasks for this mission.
(a) Completely specify the geometry of the Earth departure trajectory and the velocity change required to achieve such a departure.
(b) Completely specify the Uranus arrival and departure geometries.
(c) Determine the absolute velocity <u>vector</u> of the probe resulting from the encounter.
(d) Compute the approximate transfer time from Earth to Uranus and the phase angle required for this transfer.
(Ans. (a) $e_E = 3.23, \frac{\delta_E}{2} = 18.05^o, \Delta_E = 9.621.5$ km, $\Delta v = 7.98\frac{km}{s}$;
(b) $e_U = 1.1, \delta_U = 131.65^o, \Delta_U = 1.266 \times 10^5$ km, $v^{-}_{\infty/U} = 4.65\ \frac{km}{s}$;
(c) $\bar{v} = -3.47\,\bar{I} + 9.88\,\bar{J}\ \frac{km}{s}$; (d) $t = 16.05$ yr, $\gamma = 68.68^o$)

23.4 A space probe is parked in an 1,800 km circular Earth orbit. This probe is to perform a flyby of Saturn in order to obtain a gravity assist to continue to the outer planets. Design an interplanetary trajectory so that this probe will pass on the dark side of Saturn and result in an absolute velocity (relative to the Sun) of 13.0 $\frac{\text{km}}{\text{s}}$ after Saturn departure. Use a Hohmann transfer from Earth to Saturn, assume that all orbits lie in the ecliptic plane and that the orbits of Earth and Saturn about the Sun are circular.

(a) Completely specify the geometry of Earth departure and the velocity change required to achieve such a departure.

(b) Completely specify the Saturn arrival and departure geometries and the distance of closest approach at Saturn.

(c) Compute the approximate transfer time between Earth and Saturn, and determine the phase angle required for the transfer.

(Ans. (a) $e_E = 3.206, \frac{\delta_E}{2} = 18.175°, \Delta_E = 11{,}292.8 \text{ km}, v^+_{\infty/E} = 10.37 \frac{\text{km}}{\text{s}}$, $\Delta v_1 = 7.34 \frac{\text{km}}{\text{s}}$, (b) $e_S = 1.157,\ \delta_S = 119.577°, \Delta_S = 7.55 \times 10^5 \text{ km}$, $v^-_{\infty/S} = 5.41 \frac{\text{km}}{\text{s}} = v^+_{\infty/S}, r_S = 203{,}722 \text{ km}$ (c) $t = 6.39 \text{ yr}, \gamma = 73.49°$)

23.5 A 15,000 kg space probe is parked in a 200 km direct, circular Earth orbit. This probe is to perform a planetary flyby of Mars in order to obtain a gravity assist to continue on to the outer planets. Design an interplanetary trajectory so that the probe will pass within 300 km of the surface of Mars on the sunlit side. Use a Hohmann transfer from Earth to Mars, assume that all orbits lie in the ecliptic plane and that the orbits of Earth and Mars about the Sun are circular. All velocity changes can be considered to be impulsive and perform the following tasks for this mission

(a) completely specify the geometry of Earth departure and the velocity change required to achieve such a departure

(b) completely specify the Mars arrival and departure geometries

(c) compute the approximate transfer time between Earth and Mars and the phase angle required for this transfer

(d) determine the magnitude of the probe's heliocentric velocity after its Mars flyby.

(Ans. (a) $e_E = 1.14, \frac{\delta_E}{2} = 61.31°, \Delta_E = 25{,}104.7 \text{ km},\ \Delta v = 3.61 \frac{\text{km}}{\text{s}}$;
(b) $e_M = 1.637, \delta_M = 75.297°, \Delta_M = 7{,}933.0 \text{ km}, v^-_{\infty/M} = 2.652 \frac{\text{km}}{\text{s}}$;
(c) $t = 258.74$ days, $\gamma = 135.62°$; (d) $|\bar{V}_{SC/Sun}| = 23.60 \frac{\text{km}}{\text{s}}$)

23.6 A spacecraft travels from Earth to Jupiter on a Hohmann transfer and passes the surface of Jupiter at an altitude of 200,000 km on the sunlit side. Determine the $\Delta v$ gained by the spacecraft due to Jupiter's gravity and the semi-major axis and eccentricity of the resulting heliocentric orbit of the spacecraft.

(Ans. $\Delta v = 10.58 \frac{\text{km}}{\text{s}}$, $a = 4.791 \times 10^9 \text{km}, e = 0.845$)

23.7 A spacecraft is in a 300 km direct, circular Earth orbit and must transfer to an orbit which will pass within 1,000 km of the surface of Venus on the dark side. Considering that the planet orbits are circular and lie in the ecliptic plane, design an interplanetary trajectory to accomplish this mission. Use a Hohmann transfer from Earth to Venus and consider all velocity changes to be impulsive. Perform the following tasks

(a) completely specify the geometry of Earth departure trajectory and the velocity change required to achieve such a departure

(b) completely specify the Venus arrival and departure geometries

(c) compute the approximate transfer time between Earth and Venus, and the phase angle required for this transfer

(d) determine the inertial velocity vector of the spacecraft resulting from the encounter with Venus, and the probe's heliocentric eccentricity and semi-major axis.

(Ans. (a) $e_E = 1.105, \frac{\delta_E}{2} = 64.86^o, \Delta_E = 30{,}007.2\ \text{km}, v^+_{\infty/E} = 2.50\ \frac{\text{km}}{\text{s}}$;
(b) $e_V = 1.162, \delta_V = 118.71^o, \Delta_V = 26{,}246.1\ \text{km}, v^-_{\infty/V} = 2.72\ \frac{\text{km}}{\text{s}}$;
(c) $t = 145.91$ days, $\gamma = 234.07^o$; (d) $\bar{V} = -136.07\ \bar{I} + 33.74\ \bar{J}\frac{\text{km}}{\text{s}}$, $e = 0.044$,
$a = 124{,}824{,}521.3$ km)

23.8 An asteroid approaches Jupiter on the sunlit side with an inertial velocity (relative to the Sun) of $18.5\ \frac{\text{km}}{\text{s}}$. The eccentricity of the approach trajectory is 2.9. Assuming a circular heliocentric orbit for Jupiter, calculate the following quantities:

(a) altitude of the asteroid at its point of closest approach

(b) turning angle achieved during Jupiter passage

(c) magnitude of the inertial velocity of the spacecraft after Jupiter departure, and

(d) determine the type of resulting heliocentric orbit of the asteroid after Jupiter departure, i.e., elliptical, parabolic, or hyperbolic.

(Ans. (a) (alt) = 170.198.6 km; (b) $\delta_J = 40.34^o$; (c)$v_A = 23.2\ \frac{\text{km}}{\text{s}}$; (d) hyperbolic)

(This page was intentionally left blank.)

# Module 24: Interplanetary Trajectories - Planetary Capture

Rather than flyby a planet, it is often desirable to enter into an orbit about the planet.

Planetary capture can be achieved with a single-impulse applied to the spacecraft to reduce its speed, allowing it to be captured by the gravitational attraction of the planet.

The single-impulse is most often applied at the point of closest approach, resulting in a circular orbit about the planet at that point. Depending on the magnitude of the $\Delta\bar{v}$ applied, the spacecraft could also be captured into an elliptical orbit, entering either at the apoapsis or periapsis, or somewhere between the two.

Consider a hyperbolic approach to a planet as

As the spacecraft approaches the target planet its velocity will increase and become a maximum at the point of closest approach, $\bar{r}_p$, as shown below

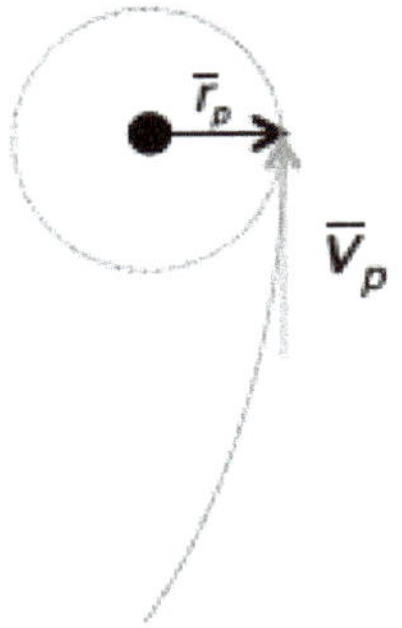

A tangential velocity change must be applied in the direction opposite to the velocity at $\bar{r}_p$ to decrease the energy of the spacecraft and allow it to be captured by the planet's gravity

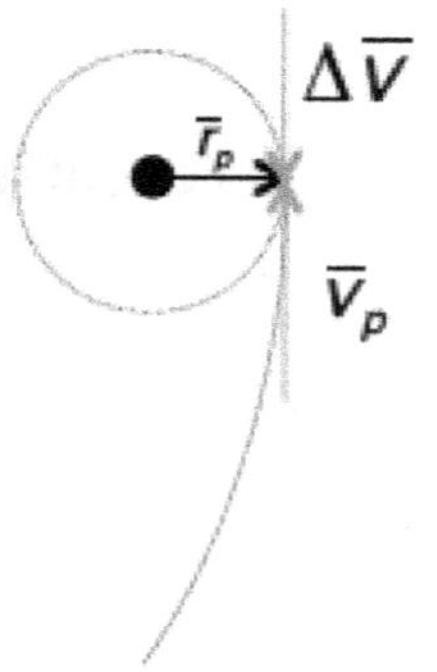


If this $\Delta\bar{v}$ reduces the spacecraft velocity to the circular velocity of the planet, $\bar{v}_c$, the spacecraft will be captured in a circlular orbit having a radius of $\bar{r}_p$, as shown below

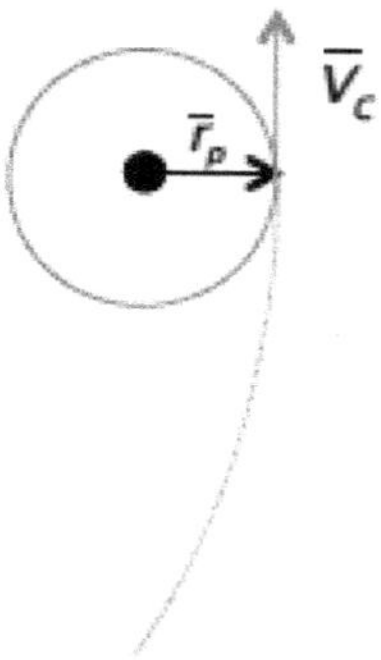


If the velocity of the spacecraft is reduced to the speed at apoapsis or at periapsis of an elliptical orbit, the spacecraft will be captured into that elliptical orbit rather than a circular orbit. Capture will occur whenever the spacecraft velocity is reduced to any speed less than the escape speed for the target planet. A variety of capture orbits can result for any $\Delta v$.

Since onboard fuel is usually limited, it is often desirable for the spacecraft to be captured by the planet with minimum fuel expenditure.

In planetary capture problems, usually the values for $r_p$ and $v_\infty^-$ are known. Therefore, the geometry of the approach trajectory can be described by the equations

$$e = \frac{r_p v_\infty^2}{\mu_{PL}} + 1$$

$$\Delta = \sqrt{\frac{\mu_{PL}^2 (e^2 - 1)}{(v_\infty^-)^4}}$$

$$\frac{\delta}{2} = \sin^{-1}\left(\frac{1}{e}\right)$$

At the point of closest approach, the spacecraft has the velocity

$$v = \sqrt{\mu_{PL}\left(\frac{2}{r_p} - \frac{1}{a}\right)}$$

An optimization problem can be solved in order to minimize Δv usage with respect to the final circular orbit radius. In doing so, an optimal hyperbolic approach trajectory can be determined for a minimum Δv capture.

Results of the optimization problem provides the optimal capture to be one having the following properties

$$\left(r_p\right)_{OPT} = \frac{2\mu}{(v_\infty^-)^2}$$

$$(\Delta v)_{MIN} = \frac{v_\infty^-}{\sqrt{2}} = \sqrt{E}$$

$$\Delta_{OPT} = \frac{\mu\sqrt{8}}{(v_\infty^-)^2}$$

$$e_{OPT} = 3$$

For optimal capture, Δ is usually very large and the spacecraft is captured in a very high altitude circular orbit.

Example 24.1

A spacecraft traveling from Earth approaches Mars on the sunlit side with a hyperbolic excess speed of 2.65 $\frac{km}{s}$. When the spacecraft reaches a distance of closest approach, $r_p$, of 5,447 km, it is to enter into an elliptical orbit about Mars having a semi-major axis of 8,332 km. Determine the following quantities
(a) the impulsive velocity change required to achieve this capture
(b) the eccentricity of the approach trajectory
(c) the aiming distance, Δ, on Mars approach
(d) the value of one-half the turning angle at Mars.

Solution:

(a) At periapsis,

$$v_p = \sqrt{\mu_M\left(\frac{2}{r_p} - \frac{1}{a}\right)} = \sqrt{(4.283 \times 10^4)\left(\frac{2}{5{,}447} - \frac{1}{8{,}332}\right)} = 3.25\ \frac{km}{s}$$

$$v_{bo} = \sqrt{(v_{\bar{\infty}})^2 + \frac{2\mu_M}{r_p}} = \sqrt{(2.65)^2 + \frac{2(4.283\times10^4)}{5{,}447}} = 4.77\ \frac{km}{s}$$

$$\Delta v = v_{bo} - v_p = 4.77 - 3.25\ \frac{km}{s}$$

$$\Rightarrow\ \Delta v = 1.52\ \frac{km}{s}$$

(b) $$e = \frac{r_p(v_{\bar{\infty}})^2}{\mu_M} + 1 = \frac{5{,}447(2.65)^2}{4.283\times10^4} + 1$$

$$\Rightarrow\ e = 1.893$$

(c) $$\Delta_M = \sqrt{\frac{\mu_M^2(e^2-1)}{(v_{\bar{\infty}})^4}} = \sqrt{\frac{(4.283\times10^4)^2[(1.893)^2-1]}{(2.65)^4}}$$

$$\Rightarrow\ \Delta_M = 9{,}802.9\ \text{km}$$

(d) $$\frac{\delta_M}{2} = \sin^{-1}\left(\frac{1}{e}\right) = \sin^{-1}\left(\frac{1}{1.893}\right)$$

$$\Rightarrow\ \frac{\delta_M}{2} = 31.9^o$$

## Example 24.2

A spacecraft from a distant planet approaches Earth with a hyperbolic excess speed of $11.90\ \frac{km}{s}$ and is to be captured in a circular orbit at an altitude of 500 km. Calculate the semi-major axis, eccentricity, and aiming distance of the approach trajectory and the velocity change required to accomplish Earth capture.

Solution:

$$r_E = R_E + (\text{alt}) = 6{,}378.1 + 500 = 6.878.1\ \text{km}$$

$$a_E = \frac{-\mu_E}{\left(v_{\infty/E}^-\right)^2} = \frac{-3.986\times10^5}{(11.90)^2}$$

$$\Longrightarrow\ a_E = -2{,}815.7\ \text{km}$$

$$e_E = \frac{r_E\left(v_{\infty/E}^-\right)^2}{\mu_E} + 1 = \frac{6{,}878.1(11.90)^2}{3.986\times10^5} + 1$$

$$\Longrightarrow\ e_E = 3.444$$

$$\Delta_E = \sqrt{\frac{\mu_E^2(e_D^2-1)}{\left(v_{\infty/E}^-\right)^4}} = \sqrt{\frac{(3.986\times10^5)^2[(3.444)^2-1]}{(11.90)^4}}$$

$$\Longrightarrow\ \Delta_E = 9{,}276.4\ \text{km}$$

$$v_c = \sqrt{\frac{\mu_E}{r_e}} = \sqrt{\frac{3.986\times10^5}{6{,}878.1}} = 7.61\ \frac{km}{s}$$

$$v_{bo} = \sqrt{\left(v_{\infty/E}^-\right)^2 + \frac{2\mu_E}{r_E}} = \sqrt{(11.90)^2 + \frac{2(3.986\times10^5)}{6.878.1}} = 16.05\ \frac{km}{s}$$

$$\Delta v = v_{bo} - v_c = 16.05 - 7.61\ \frac{km}{s}$$

$$\Longrightarrow\ \Delta v = 8.44\ \frac{km}{s}$$

Example 24.3

For the Mars approach described in Example 24.2, find the parameters of the trajectory to perform a minimum Δv capture in a circular orbit of Mars.

Solution:

$$(r_p)_{OPT} = \frac{2\mu}{(v_{\infty}^{-})^2} = \frac{2(4.283\times10^4)}{(2.65)^2}$$

$$\Rightarrow (r_p)_{OPT} = 12{,}197.9 \text{ km}$$

$$(\Delta v)_{MIN} = \frac{v_{\infty}^{-}}{\sqrt{2}} = \frac{2.65}{\sqrt{2}}$$

$$\Rightarrow (\Delta v)_{MIN} = 1.87 \ \frac{\text{km}}{\text{s}}$$

$$\Delta_{OPT} = \frac{\mu\sqrt{8}}{(v_{\infty}^{-})^2} = \frac{(4.283\times10^4)\sqrt{8}}{(2.65)^2}$$

$$\Rightarrow \Delta_{OPT} = 17{,}250.5 \text{ km}$$

$$e_{OPT} = 3$$

## Problems

24.1 Assuming that Earth and Mercury have coplanar circular orbits about the Sun, calculate the Δv requirements for a Hohmann transfer from a 150 km circular parking orbit at Earth to a 800 km capture orbit at Mercury.
(Ans. $\Delta v = 13.25 \ \frac{\text{km}}{\text{s}}$)

24.2 On a Hohmann transfer from an inferior planet to Uranus, a spacecraft approaches with a hyperbolic excess speed of $4.0 \ \frac{\text{km}}{\text{s}}$ on the sunlit side. For a point of closest approach of the hyperbolic trajectory of 25,500 km, determine the eccentricity, semi-major axis, turning angle, and aiming distance. Also find the Δv required for the spacecraft to be captured at its closest approach.
(Ans. $e_U = 1.07, a_U = -364{,}285.7$ km, $\delta_U = 138.32°, \Delta_U = 137{,}845.5$ km, $\Delta v = 6.62 \ \frac{\text{km}}{\text{s}}$)

24.3 For Problem 24.1, find the eccentricity of the resulting orbit of Mercury if the $\Delta v$ available for capture is 7.0 $\frac{km}{s}$ and the point of closest approach is also 3,240 km. (Ans. $e = 0.60$)

24.4 Prove that for an optimal capture trajectory, $e_{OPT} = 3$, and $\Delta_{OPT} = \frac{\sqrt{8}\mu_{PL}}{(v_{\infty}^{-})^2}$.

24.5 For the approach trajectory described in Example 24.1, calculate the following parameters for a minimum $\Delta v$ capture at Mars $(r_p)_{OPT}$, $\Delta_{OPT}$, and $(\Delta v)_{MIN}$.
(Ans. $(r_p)_{OPT} = 12{,}216.4$ km, $\Delta_{OPT} = 28{,}882.4$ km, $(\Delta v)_{MIN} = 1.87 \frac{km}{s}$)

24.6 In Problem 24.2, determine the parameters for a minimum $\Delta v$ capture at Uranus.
(Ans. $(r_p)_{OPT} = 72{,}375.0$ km, $\Delta_{OPT} = 102{,}353.7$ km, $(\Delta v)_{MIN} = 2.83 \frac{km}{s}$)

24.7 Assume that a space vehicle performs a Hohmann transfer from Earth to Jupiter. If it performs a flyby on the sunlit side at an altitude of 200,000 km, calculate the following quantities
(a) the velocity change required for the spacecraft to be captured at an altitude of 200,000 km
(b) the minimum velocity change for the spacecraft to be captured by Jupiter and the resulting altitude.
(Ans. (a) $\Delta v = 9.47 \frac{km}{s}$; (b) $(\Delta v)_{MIN} = 3.99 \frac{km}{s}$, (alt) = 7,886,190.7 km)

24.8 A spacecraft returning from a lunar mission approached Earth on a hyperbolic trajectory. At its closest approach its altitude is 5,000 km and its speed is 10 $\frac{km}{s}$. Retrorockets are fired 180° apart to lower the spacecraft to a 500 km circular orbit. Find the total $\Delta v$ required to accomplish this transfer.
(Ans. $\Delta v = 5.75 \frac{km}{s}$)

(This page was intentionally left blank.)

# Module 25: Satellite Perturbations

The two basic assumptions made in the solution of the two-body problem were:

- Both bodies are considered to be point masses.

- The mutual gravitational attraction of the masses are the only forces acting on the bodies. No other forces affect their motion.

Neither of those assumptions are accurate in the real world so the use of the two body equations of motion has its limitations. In actuality, both bodies are seldom point masses and there are many other forces present that create additional accelerations on the two bodies. This fact is especially true in calculating the motion of a satellite or other spacecraft.

The two-body forces are the largest forces present and establish the 'reference' orbit of the satellite. Other forces, which are smaller than the two-body forces, cause variations, or 'perturbations' to the reference orbit. The presence of these other forces acting on a satellite, their relative magnitudes, and the general solution approach for the largest of these perturbations, due to the non-spherical Earth, are addressed in this module. It is the opinion of the author that perturbations due to the non-spherical Earth, while an advanced topic, represents essential knowledge in the field of orbital mechanics and will be addressed in this textbook.

## Perturbation Forces

As a result of the other forces present, the two-body relative equations of motion must be modified to include accelerations due to these 'perturbation' forces. The acceleration model for a satellite can now be expressed as the Perturbed Relative Two-Body Problem as

$$\ddot{\bar{r}} = \frac{-\mu}{r^3}\bar{r} + \bar{a}_p \tag{19}$$

where,

$\bar{a}_p$ = perturbation accelerations, which create changes to the two-body reference orbit established by the first term on the right-side of Equation (19), the two-body accelerations

The perturbation term includes accelerations due to the presence of many other forces as described below.

$\bar{a}_{NS}$ = accelerations due to the non-spherical Earth, i.e., oblateness, or the equatorial bulge, which is the largest of the non-spherical Earth effects and is $\cong 10^{-3}$ times the two-body accelerations. These non-sphericity effects include

- pulling a satellite back towards the equatorial plane.

- rotation of the orbital plane about the ECI Z-axis.

- changing all the orbital elements in some way, the effects of which are more pronounced at lower altitudes. The value of $\Omega$ can change by several degrees per day.

$\bar{a}_G$ = accelerations due to the gravitational attraction of other bodies, i.e., Sun, Moon, and the planets, called n-body accelerations

- effects increase almost linearly with the distance from Earth.

- accelerations due to the Moon and Sun, i.e., luni-solar effects, dominate the effects of other bodies.

- significant contributions result from Venus and Jupiter for Earth satellites.

$\bar{a}_D$ = accelerations due to atmospheric drag forces (for low altitude satellites)

- drag is a surface force that's a function of mass and area.

- satellites up to several hundred km in altitude are affected.

- accelerations are velocity-dependent and are caused by atmospheric resistance.

- drag effects decrease exponentially with altitude.

- major effects of drag are a slow decrease in semi-major axis and eccentricity of an orbit.

- drag can pose severe limits to satellite lifetime but has little effect on the orbital plane.

$\bar{a}_{SRP}$ = accelerations due to solar radiation pressure (SRP) forces

- SRP is a surface force, that's a function of mass and area and doesn't vary with altitude.

- results from photons impinging on satellite surfaces.

- SRP greatly affects communication satellites with large solar panels.

- main effects of SRP are changes in eccentricity and the argument of periapsis.

$\bar{a}_{A}$ = accelerations due to Earth radiation pressure forces, or Earth albedo

- solar radiation which reflects off the Earth and onto a satellite is called albedo.

- about 30% of incoming solar radiation is reflected back.

- albedo effects on some satellites can be measurable.

$\bar{a}_{T}$ = accelerations due to forces from ocean tides and solid-Earth tides (can also include atmospheric tides).

- ocean tides cause a large change in mass distribution which must be accounted for in applications that require high levels of accuracy.

- solid-Earth tides are the deformations of Earth due to perturbing forces.

- solid-Earth tides also result from internal forces created by the motions of liquids and solids beneath Earth's surface.

$\bar{a}_{TH}$ = accelerations due to thrust forces

- function of the motor's mass-flow rate and the specific impulse of the fuel and can quickly produce significant changes in orbital elements.

- motor firings act over a finite time interval, but thrust events up to five minutes can be treated as instantaneous velocity changes.

- both thrust and mass-flow rate are functions of time and the time dependencies must be modeled.

$\bar{a}_R$ = accelerations due to relativistic forces

- speed of light is delayed slightly due to Earth's gravitational field.
- the delay creates an error of approximately 1 cm in Satellite Laser Ranging (SLR) measurements.

$\bar{a}_M$ = accelerations due to forces created by satellite interaction with Earth's magnetic field (if the satellite is metallic)

- Eddy currents can be introduced which create magnetic field forces.

$\bar{a}_C$ = accelerations due to forces from charged and uncharged particles (of solar origin)

- generally, in the form of neutral atoms and dust.

$\bar{a}_{OTHER}$ = accelerations due to other forces that may be present

The perturbation accelerations can be expressed as

$$\bar{a}_P = \bar{a}_{NS} + \bar{a}_G + \bar{a}_D + \bar{a}_{SRP} + \bar{a}_A + \bar{a}_T + \bar{a}_{TH} + \bar{a}_R + \bar{a}_M + \bar{a}_C + \bar{a}_{OTHER} \tag{20}$$

Since many other accelerations are present, the two-body model alone will not predict the motion of a satellite accurately.

The major perturbations include, $\bar{a}_{NS}$, $\bar{a}_G$, $\bar{a}_D$, and $\bar{a}_{SRP}$ while the other accelerations provide minor perturbation effects.

Consider Equation (19) as

$$\ddot{\bar{r}} = \frac{-\mu}{r^3}\bar{r} + \bar{a}_p$$

Considering the solution to this Perturbed Relative Two-Body Problem, it's true that

- there's no exact, closed-form, analytical solution exists and it's been mathematically proven that one will never exist.
- of the six integrals of the Relative Two-Body Problem, two (the Trajectory Equation and Kepler's Equation) are no longer valid.

Results of the perturbation accelerations on satellite orbits are summarized as

- the orbital elements are no longer constant and change with time.
- even though the perturbation forces are small, changes to orbital elements can be significant over time.
- two-body theory can't accurately predict the orbital elements.
- other types of solutions are required to analyze satellite motion when perturbations are included.

Relative Magnitudes of Perturbation Accelerations

The relative magnitudes of the perturbation accelerations acting on a satellite as a function of its distance from the center of Earth in shown in the figure:

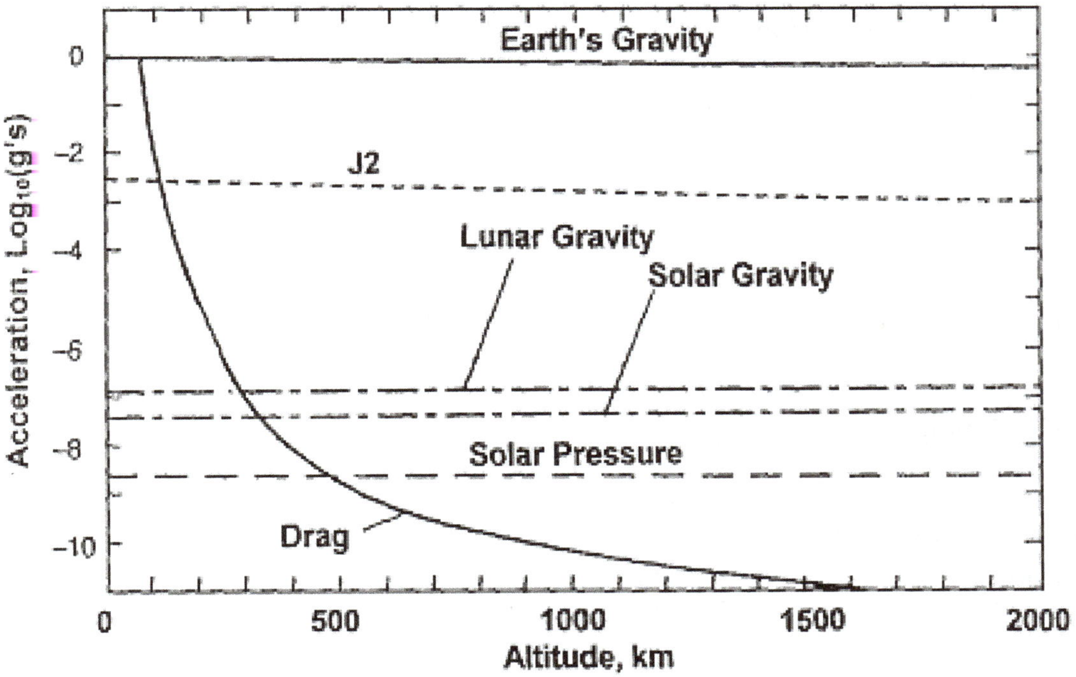


The line nearest the top of the figure, denoted as Earth's Gravity, indicates the two-body accelerations, which establishes the reference solution from which the orbit deviates due to the perturbation accelerations. A more detailed figure showing the relative perturbation accelerations on satellites is provided by Ref. [4].

The magnitude of perturbation accelerations $\left(\text{in } \frac{\text{km}}{\text{s}^2}\right)$ on a satellite in low-Earth orbit (LEO) and geosynchronous orbit (GEO) altitudes are summarized in the following table.

| Acceleration | LEO | GEO |
|---|---|---|
| Two-body | $10^{-2}$ | $10^{-3}$ |
| Non-sphericity (oblateness) | $10^{-3}$ | $10^{-8}$ |
| Sun, moon gravity | $10^{-9}$ | $10^{-8}$ |
| Other non-sphericity | $10^{-7}$ | $10^{-10}$ |
| SRP | $10^{-10}$ | $10^{-10}$ |
| Albedo | $10^{-11}$ | $10^{-12}$ |
| Solid Earth Tide | $10^{-9}$ | $10^{-12}$ |
| Venus/Jupiter gravity | $10^{-14}$ | $10^{-13}$ |
| Relativity | $10^{-11}$ | $10^{-14}$ |

As seen in this table, perturbation accelerations significantly affect satellites in LEO, primarily due to the non-sphericity of Earth.

## Solution Techniques

There are two methods commonly used to obtain solutions, or approximate solutions, to Equation (19) regardless of which or how many perturbation forces are included in the model. These two methods are known as general perturbation theory and special perturbation theory. Overviews of both methods are provided below.

General perturbation theory: replaces the equations of motion with analytical approximations that capture the character of the motion, and have the properties

- the analytical approximations produce 'general' results that hold for some limited time period and accept any initial conditions. These approximations often permit analytical integration.

- the solutions often rely on series expansions of the perturbing acceleration. The more terms included in the series increases the solution accuracy.

- the solutions are <u>qualitative</u> in nature, i.e., they describes <u>how</u> the orbital elements change with time over the time period of interest.

- solutions are more difficult to develop than numerical solutions of Equation (19) but provide more information.

Special perturbation theory (also known as <u>orbit propagation</u>): numerically integrates the equations of motion including all perturbation effects for a specific problem.

- the solution provides a specific answer for one set of initial conditions.

- the solutions can't be generalized for other problems.

- results can be very accurate, although accuracy suffers from truncation and round-off errors due to fixed computer word length.

- the numerical solutions degrade as the propagation interval gets larger.

- accuracy is a function of modeling error and its complexity.

General perturbation theory is discussed in detail for $\bar{a}_{NS}$ in this module since the non-sphericity of Earth generally has the largest effects on low-Earth satellites. As mentioned earlier, the topic of non-spherical Earth perturbations represents essential knowledge in orbital mechanics. The other perturbation forces won't be addressed in this textbook. but are discussed in significant detail in Ref. [4] and [5].

Solutions for each perturbation acceleration are treated separately, but their effects are cumulative and the principle of superposition applies.

Special perturbation theory will not be presented in this textbook as there are many quality references available that provide significant information on those solution techniques.

## Non-Spherical Earth Accelerations

A central body can be accurately modeled as a point mass only if:

- all of the mass is concentrated at a single point.

- the mass is a solid, homogeneous sphere or a thin spherical shell.

- the mass is composed of concentric homogenous layers.

Earth, Moon, and other planets are none of these, which means that they can't be accurately modeled as point masses.

Earth is <u>very</u> non-spherical and non-homogenous.

- the dominant shape of Earth is an oblate spheroid.

- it contains mountains, valleys, oceans, ice caps, mass concentrations, tectonic plates, sub-surface liquids, etc.

- Earth is actually a pulsating body with an ever changing geometry and mass distribution.

- modeling Earth as a point mass introduces many errors when used to predict near-Earth satellite motion.

The non-sphericity of Earth creates perturbation accelerations on a satellite, designated as $\bar{a}_{NS}$. The effects of this perturbation are major and create significant changes to a satellite's orbit. These changes can be summarized as

- the equatorial bulge, i.e., Earth's oblateness, pulls a satellite back towards the equatorial plane.

- these oblateness effects are $\cong 10^{-3}$ times the two-body accelerations.

- the orbital plane rotates about Z-axis of the ECI coordinate system.

- the line-of-nodes shifts up to several degrees per day.

- all orbital elements are affected in some way.

- the effects on the orbit are more pronounced at low altitudes.

Earth's shape is compared to a perfect sphere in the figure below, where the equatorial radius is 21 km larger than the radius at the poles:

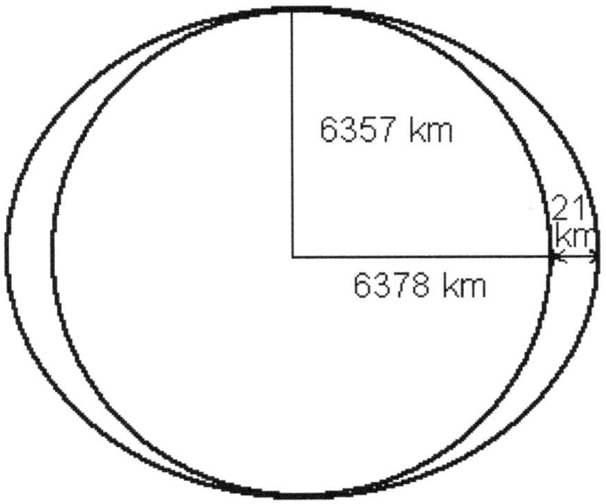


The first-order approximation of Earth's shape is generally an oblate spheroid, shown as:

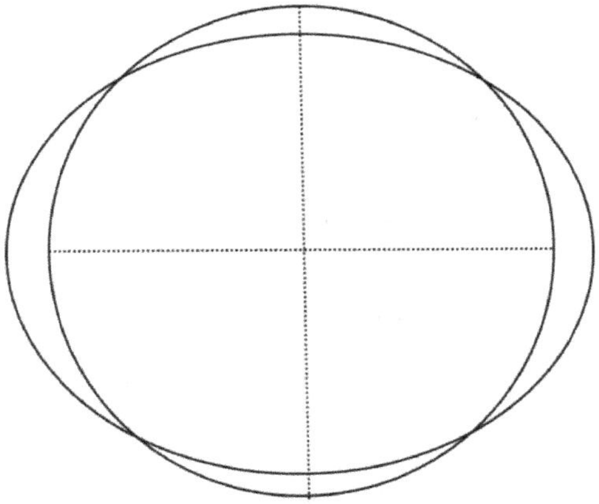

The second-order approximation of Earth's shape is similar to the shape of a pear.

Higher order models of Earth represent more complex shapes super-imposed on each other.

The equations of motion of a satellite which include the non-spherical effects of Earth, are best obtained by the development of a generalized potential function for an arbitrary non-spherical mass and computing the gradient of the potential function, shown as

$$\ddot{\bar{r}} = \bar{\nabla} U \tag{21}$$

This method is much easier than summing all the forces acting on a satellite.

The real-world problem of two arbitrarily shaped masses is extremely difficult to model mathematically. Fortunately, a satellite can generally be accurately modeled as a point mass. Consider $m_1$ to be an arbitrarily-shaped mass, and $m_2$ to be a point mass ($m_2 = 1$) located at point P as shown

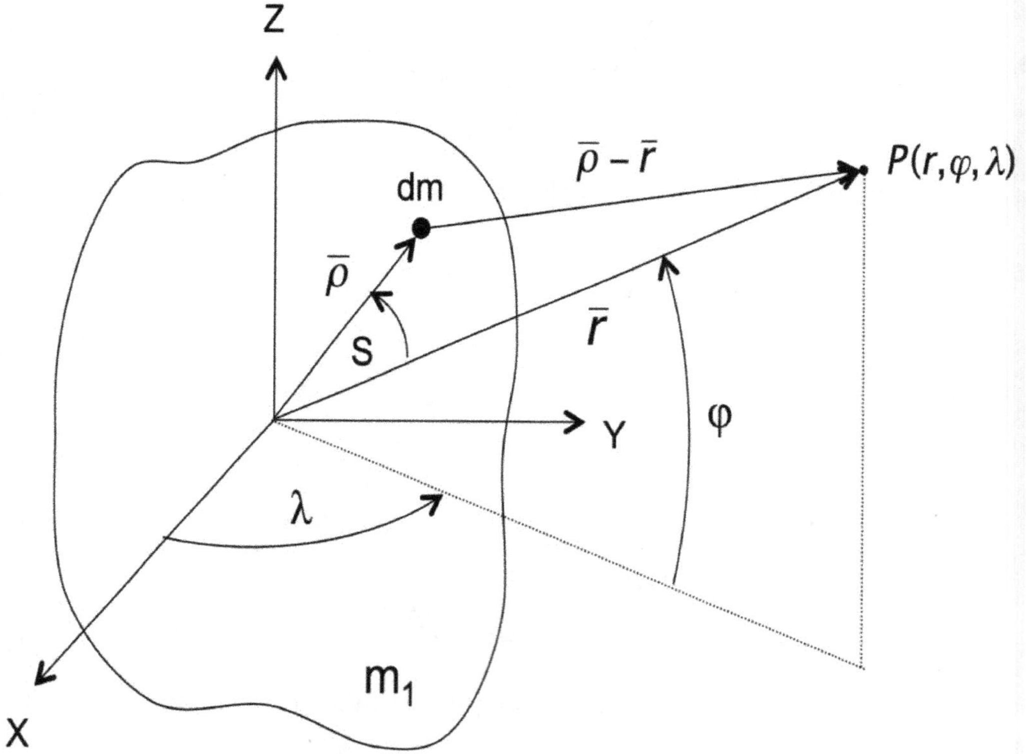


The non-uniform mass distribution of Earth not only results from its shape, but also from variations in its density and topography, etc., as mentioned above All those effects contribute to Earth's gravitational potential and can be modeled using a spherical harmonic representation of Earth's density in the ECEF coordinate system.

The gravitational potential function, i.e., geopotential, can be expressed as

$$U = \int_{m_1} \frac{G}{|\bar{r}-\bar{\rho}|} dm$$

Using spherical harmonics and the Decomposition Formula, this geopotential function can be written

$$U = \frac{\mu}{r} - \frac{\mu}{r}\sum_{n=1}^{\infty}\left[\frac{R_E}{r}\right]^n P_n(\sin\phi)J_n$$
$$+\frac{\mu}{r}\sum_{n=1}^{\infty}\sum_{m=1}^{n}\left[\frac{R_E}{r}\right]^n P_{n,m}(\sin\phi)\left[C_{n,m}\cos(m\lambda) + S_{n,m}\sin(m\lambda)\right] \quad (22)$$

where,

$n$ = degree

$m$ = order

$P_n(\sin\phi)$ = Legendre' Polynomials

$P_{n,m}(\sin\phi)$ = Legendre' Associated Functions

$J_n$ = 'zonal' harmonic coefficients = $-C_{n,0}$

$C_{n,m}, S_{n,m}$ = harmonic coefficients of Earth's gravitational field, i.e., geopotential coefficients

$\phi$ = geocentric latitude of $m_2$

$\lambda$ = east longitude

The first term on the right side of Equations (22) is the two-body, point mass potential, $U_0$.

The harmonic coefficients $C_{n,m}$, $S_{n,m}$ and $J_n$ are constants, which describe the mass distribution of mass $m_1$. If the origin of the coordinate system used to derive (22) is located at the center of mass of $m_1$, which is usually the case, it follows that

$$J_1 = C_{1,1} = S_{1,1} = 0$$

All bodies, e.g., Earth, Moon, Mars, Venus, etc., can be modeled in the same manner with the same type of geopotential function, U, but with different values of the constants.

A general formula for the calculation of the Legendre' Polynomials and the Legendre' Associated Functions, both in argument 'x', are given as

$$P_n(x) = \left(\frac{1}{2^n\, n!}\right) \frac{d^n}{dx^n} [(x^2 - 1)^n]$$

$$P_{n,m}(x) = \sqrt{(1 - x^2)^m} \frac{d^m P_n}{dx^m}$$

Sample functions through the 4th order, in argument sin ϕ, are provided by the expressions

$$P_1(\sin\phi) = \sin\phi$$

$$P_2(\sin\phi) = \frac{3}{2}\sin^2\phi - \frac{1}{2}$$

$$P_3(\sin\phi) = \frac{5}{2}\sin^3\phi - \frac{3}{2}\sin\phi$$

$$P_4(\sin\phi) = \frac{35}{8}\sin^4\phi - \frac{15}{24}\sin^2\phi + \frac{3}{8}$$

$$P_{1,1}(\sin\phi) = \cos\phi$$

$$P_{2,1}(\sin\phi) = 3\cos\phi\sin\phi$$

$$P_{2,2}(\sin\phi) = 3\cos^2\phi$$

$$P_{3,1}(\sin\phi) = \frac{15}{2}\cos\phi\sin^2\phi - \frac{3}{2}\cos\phi$$

$$P_{3,2}(\sin\phi) = 15\cos^2\phi\sin\phi$$

$$P_{3,3}(\sin\phi) = 15\cos^3\phi$$

$$P_{4,1}(\sin\phi) = \frac{35}{2}\cos\phi\sin^3\phi - \frac{15}{2}\cos\phi\sin\phi$$

$$P_{4,2}(\sin\phi) = \frac{105}{2}\cos^2\phi\sin^2\phi - \frac{15}{2}\cos^2\phi$$

$$P_{4,3}(\sin\phi) = 105\cos^3\phi\sin\phi$$

$$P_{4,4}(\sin\phi) = 105\cos^4\phi$$

The geopotential coefficients are given specific names:

$n = m \Rightarrow C_{n,m}, S_{n,m}$ are called sectoral harmonics

$n \neq m \Rightarrow C_{n,m}, S_{n,m}$ are called tesseral harmonics

Since the geopotential coefficients, $C_{n,m}$ and $S_{n,m}$, cover a range of ten or more orders of magnitude, the normalized coefficients, $\bar{C}_{n,m}$ and $\bar{S}_{n,m}$ are often used. These normalized coefficients are defined by the equation

$$\begin{Bmatrix} \bar{C}_{n,m} \\ \bar{S}_{n,m} \end{Bmatrix} = \sqrt{\frac{(n+m)!}{2(2n+1)\,(n-m)!}} \begin{bmatrix} C_{n,m} \\ S_{n,m} \end{bmatrix} \tag{23}$$

The values for the zonal, sectoral, and tesseral harmonic coefficients model the gravitational field characteristics in various regions of the non-spherical body $m_1$.

The even zonal harmonic terms, e.g., $J_2$, $J_4, J_6$, etc., model the gravitational field characteristics in regions that are symmetrical about the equator.

The odd zonal harmonic terms, e.g., $J_3$, $J_5, J_7$, etc., model the gravitational field characteristics in regions that are unsymmetrical about the equator.

Ref. [6] provides drawings that depict the modeling of the zonal harmonics through degree 5, as

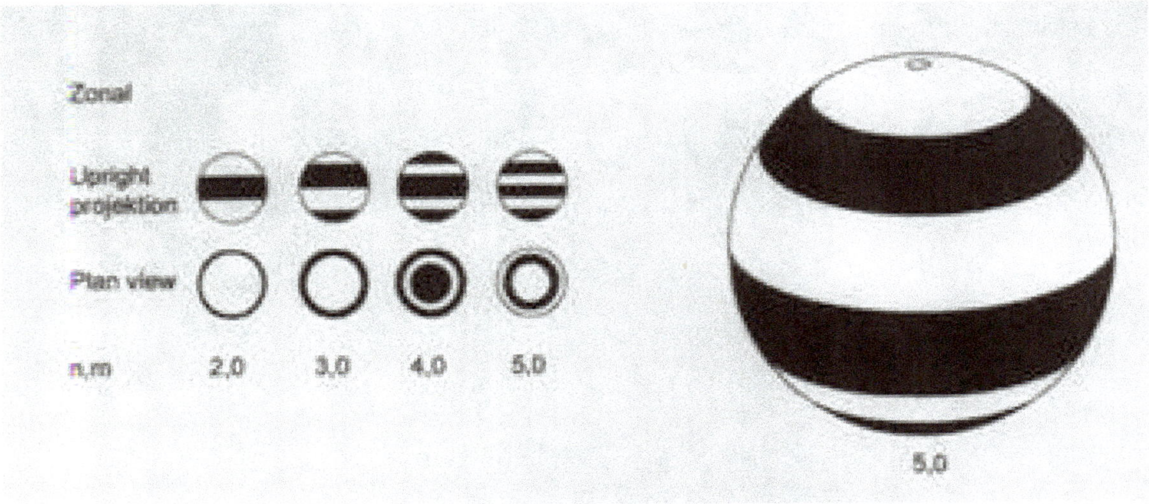

The sectoral harmonics model the gravitational field characteristics in vertical sections, i.e., ‘slices’, of Earth’s surface. Ref. [6] also provides the figure below that indicates the regions of Earth that are modeled by sectoral harmonic.

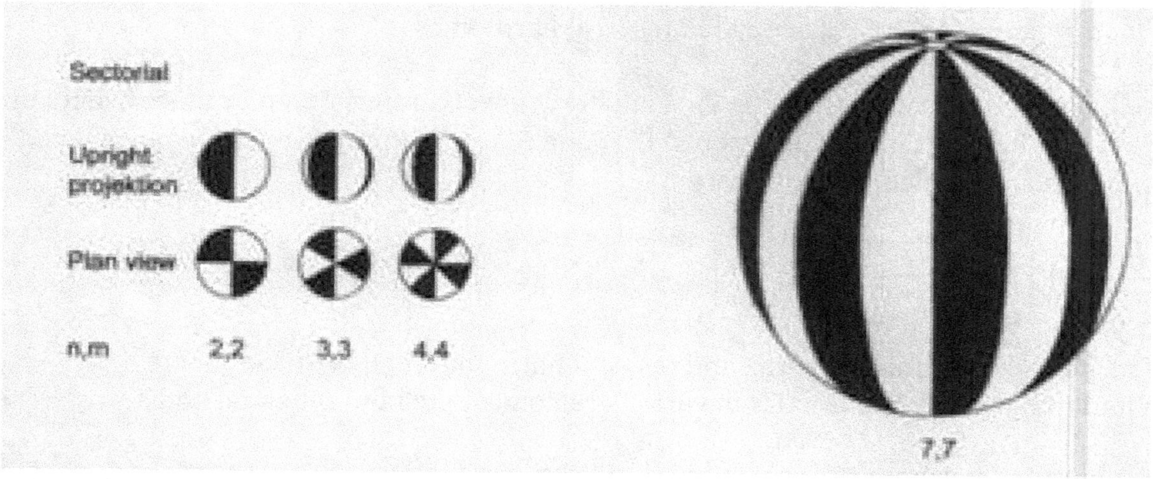


The tesseral harmonics model the gravitational field characteristics in rectangular sections, i.e., ‘tiles’ of Earth’s surface. Those tiles are shown in the following figure, also from Ref. [6].

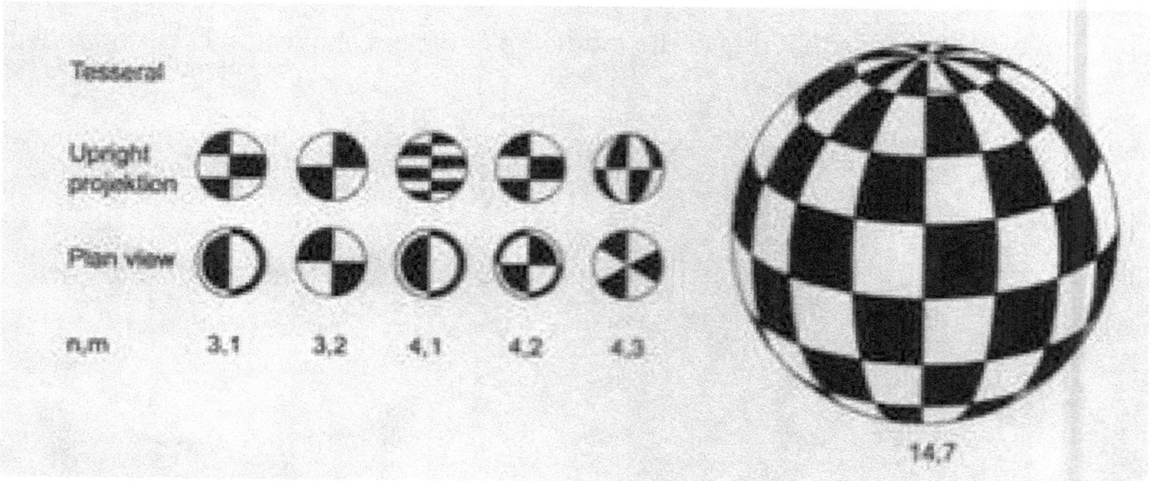

Typical sizes of the gravity field used for modeling based on the accuracy level desired and computational power available are

- low-Earth, non-scientific satellite: n, m $\sim$ 6-8 $\Rightarrow$ 36-64 coefficients.
- moderate accuracy: n, m = 36 $\Rightarrow$ 1,296 coefficients.
- scientific satellites: n, m $\sim$ 180 $\Rightarrow$ 32,400 coefficients.
- Earth Gravity Model EGM96: n, m = 360 $\Rightarrow$ 129,600 coefficients.
- Grace Gravity Model GGM02 (2005): n, m = 200 $\Rightarrow$ 40,000 coefficients.
- Grace Gravity Model GGM05 (2016): n, m = 360 $\Rightarrow$ 129,600 coefficients.

Earth constants and gravity field coefficients are determined from the analysis of satellite observational data from many different satellites, in many different orbits, taken over many years. That process is known as satellite geodesy. A recent model, Grace Gravity Model (GGM05), was published in 2016, by researchers at the Center for Space Research at the University of Texas at Austin. The Earth constants and complete set of GGM05 gravity field coefficients are available in Ref. [7]. Approximate values of the Earth constants and gravity field coefficients (non-normalized) through degree and order 2 are provided as

$$\mu = 3.986004415 \times 10^{9} \ \frac{\text{km}^3}{\text{s}^2}$$

$$R_E = 6278.13630 \ \text{km}$$

$$\omega_e = 0.7292115 \times 10^{-4} \ \frac{\text{rad}}{\text{s}}$$

$$J_2 = 1.08263566655 \times 10^{-3}$$

$$J_3 = -2.53247369133 \times 10^{-6}$$

$$J_4 = -1.61997430578 \times 10^{-6}$$

$$C_{2,1} = -2.64116010269 \times 10^{-10}$$

$$S_{2,1} = 1.80328626939 \times 10^{-9}$$

$$C_{2,2} = 1.57457641956 \times 10^{-6}$$

$$S_{2,2} = -9.03867945726 \times 10^{-7}$$

It can be seen from the values of the coefficients above that $J_2$ is the dominant effect by an order of $10^3$.

Equation (22) can be expressed as the sum of the two-body effects, $U_0$, and a <u>disturbing function</u>, R, due to non-spherical Earth perturbation effects, as

$$U = U_0 + R$$

The perturbed equations of motion, i.e., the force per unit mass, or the acceleration, acting on a satellite having coordinates (r, ϕ, λ) can be determined by taking the gradient of the potential function in spherical coordinates, as

$$\ddot{\bar{r}} = \bar{\nabla}U = \bar{\nabla}(U_0 + R) = \frac{\partial U}{\partial r}\bar{u}_r + \left(\frac{1}{r}\right)\frac{\partial U}{\partial \phi}\bar{u}_\phi + \left(\frac{1}{r\cos\phi}\right)\frac{\partial U}{\partial \lambda}\bar{u}_\lambda$$

$$\ddot{\bar{r}} = \frac{-\mu}{r^3}\bar{r} + \bar{\nabla}R \tag{24}$$

where,

$$U_0 = \frac{\mu}{r}$$

$\bar{u}_r, \bar{u}_\phi, \bar{u}_\lambda$ are unit vectors in the r, ϕ, and λ direction, respectively

and from Equation (22)

$$R = -\frac{\mu}{r}\sum_{n=1}^{\infty}\left[\frac{R_E}{r}\right]^n P_n(\sin\phi)J_n$$
$$+\frac{\mu}{r}\sum_{n=1}^{\infty}\sum_{m=1}^{n}\left[\frac{R_E}{r}\right]^n P_{n,m}(\sin\phi)\left[C_{n,m}\cos(m\lambda) + S_{n,m}\sin(m\lambda)\right] \tag{25}$$

Example 25.1

Compare the magnitude of the two-body acceleration, the acceleration due to $J_2$, and the acceleration due to $J_3$ on an Earth satellite located at a latitude of 60° and an altitude of 200 km.

Solution:

$$r = R_E + (alt) = 6{,}378.1 + 200 = 6{,}578.1 \text{ km}$$

Two-body acceleration:

$$U_0 = \frac{\mu}{r}$$

$$F_r = \frac{\partial U_0}{\partial r} = \frac{\partial\left(\frac{\mu}{r}\right)}{\partial r} = \frac{-\mu}{r^2} = \frac{-3.986\times10^5}{(6{,}578.1)^2)} = -9.212 \times 10^{-3}\,\frac{\text{km}}{\text{s}^2}$$

$$F_\phi = \left(\frac{1}{r}\right)\frac{\partial U_0}{\partial \phi} = \left(\frac{1}{r}\right)\frac{\partial\left(\frac{\mu}{r}\right)}{\partial r} = 0$$

$$F_\lambda = \left(\frac{1}{r}\right)\frac{\partial U_0}{\partial \phi} = \left(\frac{1}{r\cos\phi}\right)\frac{\partial\left(\frac{\mu}{r}\right)}{\partial \lambda} = 0$$

$$F_{TB} = \sqrt{F_r^2 + F_\phi^2 + F_\lambda^2} = \sqrt{(-9.212 \times 10^{-3})^2 + (0)^2 + (0)^2}$$

$$\Rightarrow F_{TB} = 9.212 \times 10^{-3}\,\frac{\text{km}}{\text{s}^2}$$

$J_2$ acceleration, $n = 2$, $m = 0$:

$$R_{J_2} = -\frac{\mu}{r}\left[\frac{R_E}{r}\right]^2 P_2(\sin\phi)J_2 = \frac{-\mu R_E^2 J_2}{r^3}\left(\frac{3}{2}\sin^2\phi - \frac{1}{2}\right)$$

$$F_r = \frac{\partial}{\partial r}\left(R_{J_2}\right) = \frac{\partial}{\partial r}\left[\frac{-\mu R_E^2 J_2}{r^3}\left(\frac{3}{2}\sin^2\phi - \frac{1}{2}\right)\right] = \frac{-\mu R_E^2 J_2}{r^4}(3\sin\phi\cos\phi)$$

$$F_r = \frac{-(3.986\times10^5)(6{,}378.1)^2(1.083 \text{ x } 10^{-3})}{(6{,}578.1)^4}\left(\frac{3}{2}\sin^2(60°) - \frac{1}{2}\right) = 1.758 \times 10^{-5}\,\frac{\text{km}}{\text{s}^2}$$

$$F_\phi = \left(\frac{1}{r}\right)\frac{\partial}{\partial \phi}\left(R_{J_2}\right) = \left(\frac{1}{r}\right)\frac{\partial}{\partial \phi}\left[\frac{-\mu R_E^2 J_2}{r^3}\left(\frac{3}{2}\sin^2\phi - \frac{1}{2}\right)\right] = \frac{-\mu R_E^2 J_2}{r^4}(3\sin\phi\cos\phi)$$

$$F_\phi = \frac{-(3.986\times10^5)(6{,}378.1)^2(1.083 \text{ x } 10^{-3})}{(6{,}578.1)^4}(3\sin 60° \cos 60°) = -1.218 \times 10^{-5}\,\frac{\text{km}}{\text{s}^2}$$

$$F_\lambda = \left(\frac{1}{r\cos\phi}\right)\frac{\partial}{\partial\lambda}(R_{J_2})\left(\frac{1}{r\cos\phi}\right)\frac{\partial}{\partial\lambda}\frac{\partial}{\partial\phi}\left[\frac{-\mu R_E^2 J_2}{r^3}\left(\frac{3}{2}\sin^2\phi - \frac{1}{2}\right)\right] = 0$$

$$F_{J_2} = \sqrt{F_r^2 + F_\phi^2 + F_\lambda^2} = \sqrt{(1.758 \times 10^{-5})^2 + (-1.218 \times 10^{-5})^2 + (0)^2}$$

$$\Rightarrow F_{J_2} = 2.139 \times 10^{-5}\,\frac{km}{s^2}$$

$J_3$ acceleration, $n = 3, m = 0$:

$$R_{J_3} = -\frac{\mu}{r}\left[\frac{R_E}{r}\right]^3 P_3(\sin\phi)J_3 = \frac{-\mu R_E^3 J_3}{r^4}\left(\frac{5}{2}\sin^3\phi - \frac{3}{2}\sin\phi\right)$$

$$F_r = \frac{\partial}{\partial r}(R_{J_3}) = \frac{\partial}{\partial r}\left[\frac{-\mu R_E^3 J_3}{r^4}\left(\frac{5}{2}\sin^3\phi - \frac{3}{2}\sin\phi\right)\right] = \frac{4\mu R_E^3 J_2}{r^4}\left(\frac{5}{2}\sin^3\phi - \frac{3}{2}\sin\phi\right)$$

$$F_r = \frac{4(3.986\times10^5)(6{,}378.1)^3(-2.532 \text{ x } 10^{-6})}{(6{,}578.1)^4}\left(\frac{5}{2}\sin^3(60^o) - \frac{3}{2}\sin 60^o\right)$$

$$F_r = -2.378 \times 10^{-8}\,\frac{km}{s^2}$$

$$F_\phi = \left(\frac{1}{r}\right)\frac{\partial}{\partial\phi}(R_{J_3}) = \left(\frac{1}{r}\right)\frac{\partial}{\partial\phi}\left[\frac{-\mu R_E^3 J_3}{r^4}\left(\frac{5}{2}\sin^3\phi - \frac{3}{2}\sin\phi\right)\right]$$

$$F_\phi = \frac{-\mu R_E^3 J_3}{r^5}(7.5\sin^2\phi\cos\phi - 1.5\cos 60^o)$$

$$F_\phi = \frac{-(3.986\times10^5)(6{,}378.1)^3(-2.352 \text{ x } 10^{-6})}{(6{,}578.1)^5}(7.5\sin^2 60^o\cos 60^o - 1.5\cos 60^o)$$

$$F_\phi = 3.507 \times 10^{-8}\,\frac{km}{s^2}$$

$$F_\lambda = \left(\frac{1}{r\cos\phi}\right)\frac{\partial}{\partial\lambda}(R_{J_3})\left(\frac{1}{r\cos\phi}\right)\frac{\partial}{\partial\lambda}\left[\frac{-\mu R_E^3 J_3}{r^4}\left(\frac{5}{2}\sin^3\phi - \frac{3}{2}\sin\phi\right)\right] = 0$$

$$F_{J_3} = \sqrt{F_r^2 + F_\phi^2 + F_\lambda^2} = \sqrt{(-2.3788 \times 10^{-8})^2 + (3.507 \times 10^{-8})^2 + (0)^2}$$

$$\Rightarrow F_{J_3} = 4.237 \times 10^{-8}\,\frac{km}{s^2}$$

$$F_{TB}/F_{J_2}/F_{J_3} = 9.212 \times 10^{-3}/2.139 \times 10^{-5}/4.237 \times 10^{-8}\,\frac{km}{s^2}$$

$$\Rightarrow F_{TB}/F_{J_2}/F_{J_3} = 217{,}418.7/504.84/1\,\frac{km}{s^2}$$

The form of Equation (24) is generally used for analytical studies, but no closed-form, exact analytical solution exists for the problem:

Treatment of this equation is the major problem in general perturbation theory. This is a difficult problem and its solution is more suited to special perturbation techniques.

However, an 'approximate' analytical solution can be obtained by using the method of the variation of parameters and utilizing the assumption of 'osculating orbits'.

This method gives approximate solutions for the time-rates-of-change of the orbital elements. These solutions, called the 'Lagrange Planetary Equations', provide significant information regarding the 'quality' of the solution can be obtained.

The Lagrange Planetary Equations (in terms of orbital elements with M substituted for $\nu$) are given as

$$\dot{a} = \left(\frac{2}{na}\right)\frac{\partial R}{\partial M}$$

$$\dot{e} = -\left[\frac{\sqrt{(1-e^2)}}{na^2 e}\right]\frac{\partial R}{\partial \omega} + \left[\frac{(1-e^2)}{na^2 e}\right]\frac{\partial R}{\partial M}$$

$$\dot{i} = -\left[\frac{1}{na^2\sqrt{(1-e^2)}\sin i}\right]\frac{\partial R}{\partial \Omega} + \left[\frac{\cos i}{na^2\sqrt{(1-e^2)}\sin i}\right]\frac{\partial R}{\partial \omega}$$

$$\dot{\Omega} = \left[\frac{1}{na^2\sqrt{(1-e^2)}\sin i}\right]\frac{\partial R}{\partial i}$$

$$\dot{\omega} = \left[\frac{\sqrt{(1-e^2)}}{na^2 e}\right]\frac{\partial R}{\partial e} + \left[\frac{\cos i}{na^2\sqrt{(1-e^2)}\sin i}\right]\frac{\partial R}{\partial i}$$

$$\dot{M} = n - \left(\frac{2}{na}\right)\frac{\partial R}{\partial a} - \left[\frac{(1-e^2)}{na^2 e}\right]\frac{\partial R}{\partial e}$$

where,

$$n = \sqrt{\frac{\mu}{a^3}}$$

However, to use the Lagrange Planetary Equations, the disturbing function, R, must be written in terms of the orbital elements. This is not an easy exercise and must be done using a series expansion.

As examples, expansions of the disturbing functions for $J_2$ and $J_4$ yield the following results:

$$R_{J_2} = \left(\frac{-\mu J_2 R_E^2}{a^3}\right)\frac{(0.75\sin^2 i - 0.5)(1 - 3e\cos M)}{\sqrt{(1-e^2)^3}}$$

$$R_{J_4} = \left(\frac{-\mu J_4 R_E^4}{a^5}\right)\left[\left(\frac{105}{64}\right)\sin^4 i - \left(\frac{15}{8}\right)\sin^2 i + \left(\frac{3}{8}\right)\right]\frac{(1+5e\cos M)}{\sqrt{(1-e^2)^7}}$$

Once the appropriate partial derivatives of the disturbing functions are taken, the Lagrange Planetary Equations can be evaluated to determine the time rates-of-change of the orbital elements.

Analysis of the 'quality' of these approximate solutions gives the following results:

- the changes to a, e, and i are purely periodic with zero-mean, which indicates the values of a, e, and i will oscillate about their mean values.

- the changes to $\Omega$ and $\omega$ are secular, i.e., non-periodic and unbounded), and super-imposed with a periodic (zero-mean) component. The values of $\Omega$ and $\omega$ will vary between 0° and 360°.

- the changes to M, and also $\nu$, are secular, i.e., non-periodic and unbounded, and are super-imposed with a periodic (zero-mean) component and a mixed-secular, i.e., non-periodic and unbounded) component. The values of M and $\nu$ will vary between 0° and 360°.

The disturbing function, R, is often separated into a constant part, $R_C$, and a periodic part, $R_P$, as

$$R = R_C + R_P$$

For the case of $J_2$ above, these two parts are:

$$R_C = \left(\frac{-\mu J_2 R_E^2}{a^3}\right)\frac{(0.75\sin^2 i - 0.5)}{\sqrt{(1-e^2)^3}}$$

$$R_P = \left(\frac{-\mu J_2 R_E^2}{a^3}\right)\frac{(0.75\sin^2 i - 0.5)(3e\cos M)}{\sqrt{(1-e^2)^3}}$$

Using this representation, the constant part of the disturbing function creates the secular changes in the orbital elements, while the periodic part creates the periodic changes.

Classification of Perturbations

| Perturbation | Short Period (~2 days) | Long Period (~120 days) | Secular (m = 0) |
|---|---|---|---|
| Zonals, $J_n$ | Yes | Yes | n= even |
| Sectoral $C_{n,m}, S_{n,m}$, n = m | Yes | No | No |
| Tesseral $C_{n,m}, S_{n,m}$, n ≠ m | Yes | No | No |

Secular changes are <u>by far</u> the most significant for long-term orbit prediction. Results show that

- $\Omega$, $\omega$, and $\nu$ change secularly by large amounts.

- a, e, and i change periodically by small amounts.

A Secularly Precessing Ellipse is a simple approximation used to calculate the value of the orbital elements at any time. This approximation considers <u>only</u> secular changes as

$$\dot{a} = \dot{e} = \dot{i} = 0 \qquad \dot{\Omega} \neq 0 \qquad \dot{\omega} \neq 0 \qquad \dot{M} \neq 0$$

Therefore, at any time, the orbital elements can be calculated based upon their initial values $(a_0, e_0, i_0, \Omega_0, \omega_0, M_0)$ shown by

$$a = a_0$$

$$e = e_0$$

$$i = i_0$$

$$\Omega = \Omega_0 + \dot{\Omega}\,\Delta t$$

$$\omega = \omega_0 + \dot{\omega}\,\Delta t$$

$$M = M_0 + \dot{M}\,\Delta t$$

As examples, for secular changes from $J_2$ only, the Lagrange Planetary Equations reduce to

$$\dot{\Omega} = \frac{-3J_2 n R_E^2 \cos i}{2a^2(1-e^2)^2}$$

$$\dot{\omega} = \frac{3J_2 n R_E^2 (5\cos^2 i - 1)}{4a^2(1-e^2)^2}$$

$$\dot{M} = n + \frac{3J_2 n R_E^2 (3\cos^2 i - 1)}{4a^2\sqrt{(1-e^2)^3}}$$

For secular changes from $J_4$ only, the Lagrange Planetary Equations reduce to

$$\dot{\Omega} = \frac{-15J_4 n R_E^4 \cos i\left[\left(\frac{7}{4}\right)\sin^2 i - 1\right]}{4a^2(1-e^2)^4}$$

$$\dot{\omega} = \frac{-21J_4 n R_E^4\left[\left(\frac{35}{8}\right)\sin^4 i - 5\sin^2 i + 1\right]}{8a^4(1-e^2)^4} + \frac{15J_4 n R_E^4 \cos^2 i\left[\left(\frac{7}{8}\right)\sin^2 i - 1\right]}{2a^4(1-e^2)^4}$$

$$\dot{M} = n - \frac{15J_4 n R_E^4\left[\left(\frac{35}{8}\right)\sin^4 i - 5\sin^2 i + 1\right]}{4a^4\sqrt{(1-e^2)^7}} + \frac{21J_4 n R_E^4\left[\left(\frac{35}{8}\right)\sin^4 i - 5\sin^2 i + 1\right]}{8a^4\sqrt{(1-e^2)^5}}$$

Similar equations can be obtained for all even zonals and the effects are cumulative.

Effects of secular changes on the orbital plane show

Longitude of the ascending node, $\Omega$:

- $0^o < i < 90^o \Rightarrow \dot{\Omega} < 0 \Rightarrow$ nodal regression ($\Omega$ gets smaller, CW rotation)
- $90^o < i < 180^o \Rightarrow \dot{\Omega} > 0 \Rightarrow$ nodal advance ($\Omega$ gets larger, CCW rotation)
- $i = 90^o \Rightarrow \dot{\Omega} = 0 \Rightarrow$ no change in $\Omega$

Argument of periapsis, $\omega$:

- $63.4^o < i < 116.6^o \Rightarrow \dot{\omega} < 0 \Rightarrow$ perigee regression ($\omega$ gets smaller, CW rotation)
- $0^o \leq i < 63.4^o$ and $116{,}6^o < i \leq 180^o \Rightarrow \dot{\omega} > 0 \Rightarrow$ perigee advance ($\omega$ gets larger, CCW rotation)
- $i = 63.4^o, 116.6^o$ = 'critical inclination' $\Rightarrow \dot{\omega} = 0 \Rightarrow$ no change in $\omega$

Example of secular changes:

LEO:

- $T = 90 \text{ min} \Longrightarrow a = 6{,}652.555 \text{ km}$

- $e_0 = 0.025, i_0 = 28.5^o, \Omega_0 = 60.0^o, \omega_0 = 10.0^o, M_0 = 75.0^o$

$\Longrightarrow \dot{\Omega} = -7.637 \frac{o}{day}, \; \dot{\omega} = 12.419 \frac{o}{day}, \; \dot{M} = 5{,}765.667 \frac{o}{day} \; (5.666 \frac{o}{day}$ due to $J_2)$

- After one week: $\Longrightarrow \Omega = 6.541^o, \; \omega = 96.933^o, \; M = 114.669^o$

Geosynchronous Orbit:

- $T = 24 \text{ hr} \Longrightarrow a = 42{,}241.1 \text{ km}$

- $e_0 = 0.025, i_0 = 28.5^o, \Omega_0 = 60.0^o, \omega_0 = 10.0^o, M_0 = 75.0^o$

$\Longrightarrow \dot{\Omega} = -0.012 \frac{o}{day}, \; \dot{\omega} = 0.019 \frac{o}{day}, \; \dot{M} = 360.009 \frac{o}{day} \; (0.009 \frac{o}{day}$ due to $J_2)$

- After one week: $\Longrightarrow \Omega = 59.916^o, \; \omega = 10.133^o, \; M = 75.062^o$

Geostationary Orbit:

- $T = 24 \text{ hr} \Longrightarrow a = 42{,}241.1 \text{ km}$

- $e_0 = 0.000, i_0 = 0.0^o, \Omega_0 = 60.0^o, \omega_0 = 10.0^o, M_0 = 75.0^o$

$\Longrightarrow \dot{\Omega} = -0.014 \frac{o}{day}, \; \dot{\omega} = 0.027 \frac{o}{day}, \; \dot{M} = 360.014 \frac{o}{day} \; (0.014 \frac{o}{day}$ due to $J_2)$

- After one week: $\Longrightarrow \Omega = 59.902^o, \; \omega = 10.186^o, \; M = 75.096^o$

- Geostationary circular orbits eliminate most periodic effects.

Earth's oblateness, or equatorial bulge, exerts a force that pulls the satellite back to the equatorial plane and tries to align the orbital plane with the equator.

Example 25.2

Use the Lagrange Planetary Equations to compute the approximate values of the orbital elements of the satellite given below after a period of 14 days in orbit. Consider secular perturbations due to $J_2$ only.

$a = 6{,}652.555663$ km  $e = 0.025$  $i = 55^o$  $\Omega = 35^o$  $\omega = 45^o$  $M = 65^o$

Solution:

$$n = \sqrt{\frac{\mu}{a^3}} = \sqrt{\frac{3.986\times10^5}{(6{,}652.555663)^3}} = 1.163552 \times 10^{-3}\,\frac{\text{rad}}{\text{s}}$$

$$\dot{\Omega} = \frac{-3J_2nR_E^2\cos i}{2a^2(1-e^2)^2} = \frac{-3(1.0826\times10^{-3})\,(1.163552\times10^{-3})(6{,}378.1)^2\cos 55^o}{2(6{,}652.555663)^2[1-(0.025)^2]^2}$$

$$\Rightarrow \dot{\Omega} = -9.974477 \times 10^{-7}\,\frac{\text{rad}}{\text{s}} = -4.937721\frac{^o}{\text{day}}$$

$$\dot{\omega} = \frac{3J_2nR_E^2(5\cos^2 i-1)}{4a^2(1-e^2)^2} = \frac{3(1.0826\times10^{-3})\,(1.163552\times10^{-3})(6{,}378.1)^2(5\cos^2 55^o-1)}{4(6{,}652.555663)^2[1-(0.025)^2]^2}$$

$$\Rightarrow \dot{\omega} = 5.607828 \times 10^{-7}\,\frac{\text{rad}}{\text{s}} = 2.776074\frac{^o}{\text{day}}$$

$$\dot{M} = n + \frac{3J_2nR_E^2(3\cos^2 i-1)}{4a^2\sqrt{(1-e^2)^3}}$$

$$\dot{M} = 1.163552 \times 10^{-3} + \frac{3(1.0826\times10^{-3})\,(1.163552\times10^{-3})(6{,}378.1)^2(3\cos^2 55^o-1)}{4(6{,}652.555663)^2\sqrt{[1-(0.025)^2]^3}}$$

$$\Rightarrow \dot{M} = 1.163541 \times 10^{-3}\,\frac{\text{rad}}{\text{s}} = 5{,}759.9414\frac{^o}{\text{day}}$$

after 14 days,

$$a = a_0 = 6{,}652.555663 \text{ km}$$

$$e = e_0 = 0.025$$

$$i = i_0 = 55^o$$

$$\Omega = \Omega_0 + \dot{\Omega}\,\Delta t = 35 - (4.937721)(14) = -34.1281^o$$

$$\Rightarrow \Omega = 325.87^o$$

$$\omega = \omega_0 + \dot{\omega}\,\Delta t = 45 + 2.776074(14) = 83.8650^{o}$$

$$\Rightarrow \ \omega = 83.87^{o}$$

$$M = M_0 + \dot{M}\,\Delta t = 65 + 5{,}759.9414(14) = 80{,}704.1798^{o}$$

$$\Rightarrow \ M = 64.18^{o}$$

Example 25.3

A reconnaissance satellite is to be placed in an orbit having $e = 0.3$ and $i = 80^{o}$. To satisfy mission objectives, the maximum change in the eccentricity of the orbit due to $J_2$ must be equal to $4.0 \times 10^{-7}/\text{sec}$. Determine the required semi-major axis of the orbit assuming that the rate-of-change of the eccentricity is constant.

Solution:

$$R_P = \left(\frac{-\mu J_2 R_E^2}{a^3}\right)\frac{(0.75\sin^2 i - 0.5)(3e\cos M)}{\sqrt{(1-e^2)^3}}$$

$$\frac{\partial R_P}{\partial M} = \left(\frac{\mu J_2 R_E^2}{a^3}\right)\frac{(0.75\sin^2 i - 0.5)(3e\sin M)}{\sqrt{(1-e^2)^3}}$$

$$\frac{\partial R_P}{\partial M} = 0$$

$$\dot{e} = -\left[\frac{\sqrt{(1-e^2)}}{na^2 e}\right]\frac{\partial R}{\partial \omega} + \left[\frac{(1-e^2)}{na^2 e}\right]\frac{\partial R}{\partial M} = \left[\frac{(1-e^2)}{na^2 e}\right]\frac{\partial R}{\partial M}$$

$$\dot{e} = \left[\frac{(1-e^2)}{na^2 e}\right]\left(\frac{\mu J_2 R_E^2}{a^3}\right)\frac{(0.75\sin^2 i - 0.5)(3e\sin M)}{\sqrt{(1-e^2)^3}} = \frac{3\sqrt{\mu}\,J_2 R_E^2}{\sqrt{(1-e^2)a^7}}(0.75\sin^2 i - 0.5)\sin M$$

$$a = \sqrt[7]{\left\{\frac{3\sqrt{\mu}\,J_2 R_E^2}{\sqrt{(1-e^2)}(\dot{e})}(0.75\sin^2 i - 0.5)\sin M\right\}^2}$$

for maximum change, $\sin M = 1$,

$$a = \sqrt[7]{\left\{\frac{3\sqrt{(3.986\times10^5)}\,(1.0826\times10^{-3})(6{,}378.1)^2}{\sqrt{[1-(0.3)^2]}(4.0\times10^{-7})}(0.75\sin^2 80^{o} - 0.5)(1)\right\}^2}$$

$$\Rightarrow \ a = 8{,}189.8 \text{ km}$$

Problems:

25.1 Using the spherical harmonic representation of the gravitational potential function, write out the terms for the potential function including terms through degree and order 4, i.e., $n = m = 4$. Include expressions for all Legendre polynomials and Legendre Associated Functions required for the calculations.

25,2 Compare the magnitude of the acceleration due to $J_2$ and the acceleration due to $J_4$ on an Earth satellite located at a latitude of 55° and an altitude of 210 km.
(Ans. $\frac{F_{J2}}{F_{J3}} = \frac{681}{1}$)

25.3 Use the Lagrange Planetary Equations to compute the approximate values of the orbital elements of the satellite given below after a period of 14 days in orbit. Consider secular perturbations due to $J_2$ only.

$a = 6{,}652.555663$ km $\quad e = 0.03 \quad i = 45^o \quad \Omega = 30^o \quad \omega = 40^o \quad M = 75^o$

(Ans. $a = 6{,}652.555663$ km, $e = 0.03, i = 45^o$, $\omega = 130.44^o, \Omega = 304.754^o$, $M = 106.080^o$)

25.4 A satellite is to be placed in an orbit having $e = 0.35$ and $i = 75^o$. In order to satisfy mission objectives, the maximum change in the eccentricity of the orbit due to $J_2$ must be equal to $5.0 \times 10^{-7}$/sec. Compute the required semi-major axis of the orbit assuming that the rate-of-change of the eccentricity is constant.
(Ans. $a = 7{,}443.2$ km)

25.5 Find the latitude where the radial force on a satellite due to $J_2$ is equal to zero.
(Ans. $\phi = \pm 35.26^o$)

25.6 Consider the case where the origin of the relative coordinate system does not coincide with the center -of-mass of $m_1$, such that $U_1 = \frac{\mu}{r^3}\left[x\bar{\xi} + y\bar{\eta} + z\bar{\zeta}\right]$ in Cartesian coordinates. If an Earth satellite is located at $\bar{r} = 7{,}000\,\bar{\imath}$ km and the acceleration on the satellite due to $U_1$ is $\bar{a} = 1.627 \times 10^{-5}\bar{\imath} + 4.648 \times 10^{-6}\bar{k}\frac{km}{s^2}$, calculate the location of the center-of-mass, $\bar{\rho}_{CM} = \bar{\xi}\,\bar{\imath} + \bar{\eta}\,\bar{\jmath} + \bar{\zeta}\,\bar{k}$.
(Ans. $\bar{\rho}_{CM} = -7.0\,\bar{\imath} + 4.0\,\bar{k}$ km)

25.7 Determine the relationship between the terms in the Cartesian representation of $U_1$ (given in Problem 25.6) and the terms in the spherical representation of $U_1$. That is, find $\bar{\xi}$, $\bar{\eta}$, and $\bar{\zeta}$ in terms of $J_1$, $C_{1,1}$ and $S_{1,1}$.
(Ans. $\bar{\xi} = C_{1,1}R_E, \bar{\eta} = S_{1,1}R_E,\ \bar{\zeta} = -J_1 R_E$)

25.8 Use the Lagrange Planetary Equations to compute the approximate values of the orbital elements of the satellite given below after a period of 7 days in orbit. Consider secular perturbations due to $J_4$ only.

$a = 6{,}652.555663$ km  $e = 0.04$  $i = 30^o$  $\Omega = 10^o$  $\omega = 20^o$  $\nu = 30^o$

(Ans. $a = 6{,}652.555663$ km, $e = 0.04$, $i = 30^o$, $\Omega = 9.90^o$, $\omega = 19.96^o$, $\nu = 30.00^o$)

(This page was intentionally left blank.)

# References


[1] Escobal, P. R., *Methods of Orbit Determination*, John Wiley & Sons, New York, 1965.

[2] https://spaceflight.nasa.gov/realdata/sightings/SSapplications/Post/JavaSSOP/SSOP_Help/tle_def.html, January 2021.

[3] https://www.faa.gov/about/office_org/headquarters_offices/avs/offices/aam/cami/library/online_libraries/aerospace_medicine/tutorial/media/III.4.1.4_Describing_Orbits.pdf, January 2021.

[4] Montenbruck, O., and Gill, E., *Satellite Orbits: Models, Methods, and Applications*, Springer-Verlag, Berlin Heidelberg, 2000.

[5] Vallado, D. A., *Fundamentals of Astrodynamics and Applications*, Second Edition, Microcosm Press and Kluwer Academic Publishers, El Segundo, CA and Dordrecht, The Netherlands, 2001.

[6] Walter, U., *Astronautics*, Wiley-VCH, Weinham, 1998.

[7] Grace Recovery and Climate Experiment, GGM05, http://www2.csr.utexas.edu/grace/gravity/, 2016.

(This page was intentionally left blank.)

# Appendix A: Astronomical Data

The table below provides physical data for the Sun, the planets, and the moon.

| Body | Mass (kg) | $\mu$ $(\frac{km^3}{s^2})$ | Radius (km) | Semi-Major Axis (km) |
|---|---|---|---|---|
| Sun | $1.989 \times 10^{30}$ | $1.327 \times 10^{11}$ | 696,000 | - |
| Moon | $7.348 \times 10^{22}$ | $4.903 \times 10^{3}$ | 1,737 | $3.844 \times 10^{5}$ |
| Mercury | $3.302 \times 10^{23}$ | $2.203 \times 10^{4}$ | 2,440 | $5.791 \times 10^{7}$ |
| Venus | $4.869 \times 10^{24}$ | $3.249 \times 10^{5}$ | 6,052 | $1.082 \times 10^{8}$ |
| Earth | $5.974 \times 10^{24}$ | $3.986 \times 10^{5}$ | 6,378.1 | $1.496 \times 10^{8}$ |
| Mars | $6.419 \times 10^{23}$ | $4.283 \times 10^{4}$ | 3,396 | $2.279 \times 10^{8}$ |
| Jupiter | $1.899 \times 10^{27}$ | $1.267 \times 10^{8}$ | 71,490 | $7.786 \times 10^{8}$ |
| Saturn | $5.685 \times 10^{26}$ | $3.793 \times 10^{7}$ | 60,270 | $1.433 \times 10^{9}$ |
| Uranus | $8.683 \times 10^{25}$ | $5.794 \times 10^{6}$ | 25,560 | $2.872 \times 10^{9}$ |
| Neptune | $1.024 \times 10^{26}$ | $6.835 \times 10^{6}$ | 24,760 | $4.495 \times 10^{9}$ |
| Pluto | $1.25\times 10^{22}$ | $8.30 \times 10^{2}$ | 1,195 | $5.870 \times 10^{8}$ |

Other useful constants are provided below

Constant of Universal Gravitation, $G = 6.6742 \text{ x } 10^{-20} \frac{\text{km}^3}{\text{kg}-\text{s}^2}$

Earth rotation rate, $\omega_e = 7.292 \text{ x } 10^{-3} \frac{\text{rad}}{\text{s}} = 2.507 \text{ x } 10^{-1} \frac{\text{o}}{\text{min}}$

Obliquity of the ecliptic, $\varepsilon = 23.442405^{\circ}$

Orbital data for each body is provided in the table below.

| Body | Eccentricity | Inclination | Orbital Period | Rotation Period |
|---|---|---|---|---|
| Sun | - | - | - | 25.38 d |
| Moon | 0.0549 | $5.145^o$ | 27.322 d | 27.32 d |
| Mercury | 0.2056 | $7.00^o$ | 87.97 d | 58.65 d |
| Venus | 0.0067 | $3.39^o$ | 224.7 d | 243 d* |
| Earth | 0.0167 | $0.00^o$ | 365.256 d | 23.9345 h |
| Mars | 0.0935 | $1.850^o$ | 1.881 y | 24.62 h |
| Jupiter | 0.0489 | $1.304^o$ | 11.86 y | 9.925 h |
| Saturn | 0.0565 | $2.485^o$ | 29.46 y | 10.66 h |
| Uranus | 0.0457 | $0.772^o$ | 84.01 y | 17.24 h* |
| Neptune | 0.0113 | $1.769^o$ | 164.8 y | 16.11 h |
| Pluto | 0.2444 | $17.16^o$ | 247.7 y | 6.37 d* |

* denotes retrograde rotation

# INDEX

(This page was intentionally left blank.)

www.ingramcontent.com/pod-product-compliance
Lightning Source LLC
Chambersburg PA
CBHW080930220326
41598CB00034B/5740
* 9 7 9 8 9 8 6 6 1 8 3 2 6 *